"I insist upon being among those who have gone too far. In defense of freedom and the marvelous, there is but one watchword: STOP AT NOTHING."
—*Franklin Rosemont
of the Chicago Surrealist Group*

The cover was illustrated by Kelley Hensing

DEAR READER,

I USED TO THINK STEAMPUNK WAS A GENRE OF FICTION—a good one, of course, full of clanking machines and fantastical despots, but still basically just a genre of fiction. Like the best fiction, as Cory Doctorow discusses in this issue, steampunk was socially relevant, applicable to the real world as parable. But then, I don't know, six, maybe seven years ago, steampunk became more than that. It became something physical. It came off the pages and into our lives. It informs the ways we interact with machinery, one another, and society at large.

In much the same way, I used to think surrealism was a style of art. Melting clocks or whatever. But around the same time I started to really get the hang of what steampunk had to offer, a friend of mine—Kate Khatib, who helped edit the first several issues of this magazine—told me that everything I knew about surrealism was wrong. Surrealism was not an art movement, it was a cultural movement that happened to create art as a byproduct of life. And what lives they led!

Product is the byproduct of action. The act of writing is more important than having written, because the act of living is more important than having lived. What matters is the doing. And these surrealist ideas resonate with the steampunk in me. We glorify the tinkerer for tinkering, not just for having produced. Our machines show their seams and rivets proudly because the way we built them matters to us more, perhaps, than their actual functions.

But surrealism goes deeper than that. André Breton, the founder of surrealism, said that above everything else surrealism was a revolutionary movement. A movement that aimed to break people out of the status quo, to show people the real possibilities of living. To replace drudgery with the marvelous.

Enter the Red Fork Empire. For those readers who remain unfamiliar with the RFE, it is a steampunk empire of sorts—ruled by a madman emperor—that has declared war upon the dull. They make their appearance at conventions throughout the northeast United States, entertaining themselves and others. It's a LARP thing, yes, but I'll make the case that it touches on something more than that. The battle against the dull is the battle for the marvelous.

So I ask you, I ask us: how can we do more with steampunk? Or, more importantly, how can we do more with our lives? How can we confront and destroy banality in all its guises?

There's a sense of a transient community that forms at each steampunk event and weaves the conventions together, a community that steps outside the scripted existence that has been set before us by mainstream society. Let's push it. Let's take it further. Let's get it outside the hotel walls and into the rest of our lives.

Let's tear down the walls between costume and day-to-day wear. Let's tear down the walls of polite society and *be punks*. Just, you know, the sort that drink tea and the sort that have good manners when dealing with those deserving of respect.

Let's intervene in the machinery of the dull. Let's sabotage the industry that annihilates our world as surely as it annihilates our imaginations. Let's détourn the status quo into something that suits us and encourage others to do the same.

This issue of *SteamPunk Magazine*, quite possibly the last with myself as editor, is themed around two things: steampunk as surrealism, steampunk as community. We've interviewed Eric Larson, the man behind TeslaCon; convention favorites Frenchy & the Punk; and the steampunk crew The Vagabonds. We present the unconsciously surrealist Red Fork Empire and their loyal opposition, The Chaos Front, who agrees with their fight against the dull but not their authoritarian structure. And we reproduce an essay by Franklin Rosemont, a co-founder of the Chicago Surrealist Group, about the marvelous.

There's more in here than that, of course. Interviews with authors Cherie Priest and Cory Doctorow. Victorian fashion. Artificial intelligence. Edison's ghost-hunting machine. Dieselpunk, class warfare, paper dolls. How to make hydrogen-filled condom zeppelins. How to sew a fancy pouch. And fiction—amazing fiction from after the apocalypse to 19th century Italy.

—*Margaret Killjoy*

CONTENTS

FICTION:

Scarabs 38
   Katie Casey

Back to Matese 54
   reginazabo

The Iron Garden 72
   Erin Searles

Hell's Dinner Plate 84
   The Catastrophone Orchestra

The Scouts of the Pyre, Part II 106
   David Z. Morris

FEATURES:

The Chatelaine 4
   Joanna Church

Séance Through Science 8
   John Reppion

A Dieselpunk Cookbook 12
   Larry Amyett, Jr.

Steam-Punk-O-Matic 20
   Dr. Geof

Creative Machines 22
   Douglas Summers-Stay

The New Orleans General Strike of 1892 Laughs at Your Shitty Attempts to Divide the Working Class With Racism 32
   Miriam Roček

A Healthy Alternative to Fascism in Fashion 96
   Miriam Roček

Notes from the Bucket Shop 102
   Professor A. Calamity

EDITORIAL:

Punking the Past 70
   James H. Carrott

Freedom of the Marvelous 94
   Franklin Rosemont

Join the War Against the Dull 116
   The Red Fork Empire & the Chaos Front

DO-IT-YOURSELF:

Inventing Steampunk Fashion With Paper Dolls 42
   Charlotte Whatley

Sew Yourself a Pouch 48
   E.M. Johnson

Home Brewing Miniature Vegan Airships 78
   Professor Offlogic

INTERVIEWS:

Cory Doctorow 6
Cherie Priest 18
Frenchy & The Punk 28
The Vagabonds 44
Eric Larson 64
BB Blackdog 104

WHAT IS STEAMPUNK? WHAT IS STEAMPUNK MAGAZINE?

THE TERM "STEAMPUNK" WAS COINED TO refer to a branch of "cyberpunk" fiction that concerned itself with Victorian-era technology. However, steampunk is more than that. Steampunk is a burgeoning subculture that pays respect to the visceral nature of antiquated technology.

It's about "steam," as in steam engines, and it's about "punk," as in counter-culture. For an excellent manifesto, refer to the first article in our first issue, "What Then, is Steampunk?"

SteamPunk Magazine is a print publication that comes out erratically. Full quality print PDFs of it are available for free download from our website, and we keep the cost of the print magazine as low as possible. Back issues can be downloaded and the first 7 issues have been anthologized in a reader available from our publisher, Combustion Books. All work on the magazine, including articles, editing, illustration, and layout, is done by volunteers and contributors. To see how you can get involved, see page 118.

WWW.STEAMPUNKMAGAZINE.COM
COLLECTIVE@STEAMPUNKMAGAZINE.COM

VICTORIAN ACCESSORIES: THE CHATELAINE

By Joanna Church
Illustration by Cécile Matthey

Though the correct silhouette is important when creating a 19th (or any) century look, it is only the beginning. Accessorizing is not simply fun, it is sometimes essential—after all, fashion is in the details. Whether the aim of your ensemble is "authentic" or "inspired," accessories can make or break both your outfit and your mood.

Take, for example, the chatelaine: a metal clasp or hook to hang from your belt, from which hangs useful tools and objects. Though the name may conjure images of a dour housekeeper carrying a bunch of iron keys at her waist (and rightly so, since that's where the name comes from), don't imagine that the over-decorated Victorians left it at that. After all, what's useful in plain form is twice as good taken a few steps over the top! The rings of keys evolved into elaborately engraved belt clasps hung with fanciful (and interchangeable) devices... it's like the era was *asking* to be turned into an exuberant and creative subculture.

Technically, a chatelaine is defined as both the mistress of a chateau or large estate, and "a clasp or hook for a watch, purse, or a bunch of keys;" some older sources use the term to describe the chain or belt as well as the clasp. The lady of the house (or her delegate, the housekeeper) kept the keys about her person at all times, hence the migration of the name from human to belt. The fashionable, rather than simply utilitarian, chatelaine (known at the time as an equipage) appeared in the 18th century but fell out of favor around the 1790s, as diaphanous, Empire-waisted gowns did not lend themselves to its use. When natural-waisted, voluminous skirts reappeared several decades later so did the trusty chatelaine, and it stayed fashionable (though not ubiquitous) into the 20th century.

In many ways the chatelaine is the ultimate accessory, as it allows you to combine as many useful (or useless) implements as you choose into one article of convenience, without ruining the line of your skirts with bulky pockets. In addition, the style and design can vary without disturbing the functionality—you can opt for a romantic extravaganza, a quiet classical look, or anything in between. Do your plans for the day include plying your sewing needle? Nursing an infant? Buttoning or unbuttoning your gloves? Falling into a faint? Do you anticipate needing your spectacles, a notebook, some whiskey, a timepiece, face powder, matches? The properly equipped chatelaine is there for you. And those are only a few of the vintage, late 19th century options available; no doubt your own fertile imagination and crafty/recycling skills will provide more. By changing out the tools on your chatelaine, you can go from Angel in the House to Raging Suffragist in no time flat. As long as your belt stays up—and you can stand the level of clanking and rattling created by your accumulated accoutrements—you're good to go.

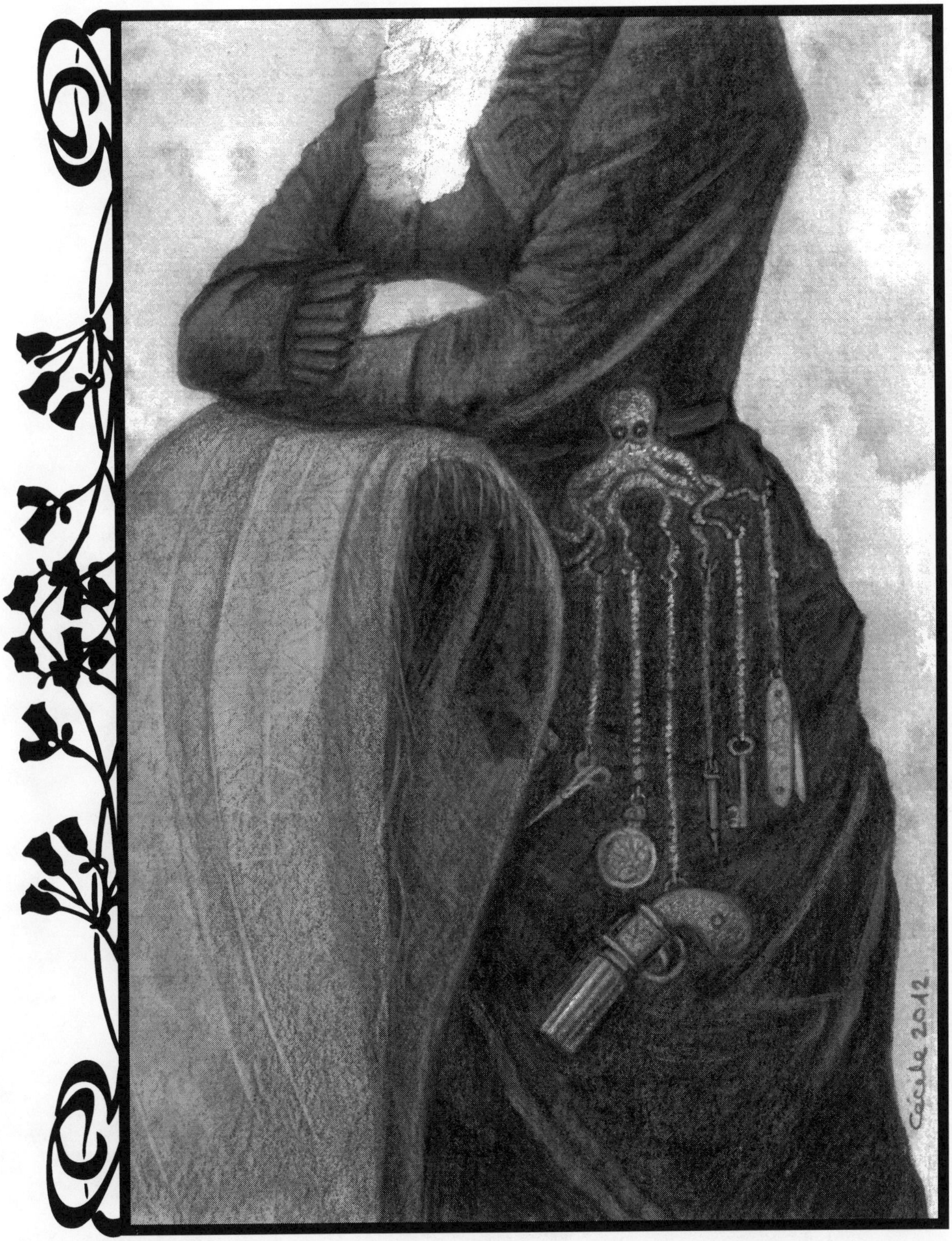
Cécile 2012

An interview by Margaret Killjoy
Illustration by Sergei Tuterov

CORY DOCTOROW

Cory Doctorow is a renowned novelist and social commenter whose work explores still-developing technologies and how they affect society. He's also, more pertinently to this magazine, one of the internet's first and most-influential bloggers of all things steampunk. It was his postings on his blog Boing Boing, in fact, that brought this magazine the audience it has. I was excited to get the chance to hear his thoughts on steampunk, maker culture, and the social effects of writing.

STEAMPUNK MAGAZINE: *You've been one of the first bloggers to pick up on steampunk, almost ten years ago now. What drew you to it, and what kept you interested?*

CORY DOCTOROW: Well, I've read steampunk fiction for a long time. I read *The Difference Engine* [by William Gibson and Bruce Sterling] when it came out. I read KW Jeter when he started writing. So of course when steampunk got its second go, it was pretty exciting to see this obscure literary subgenre that I quite enjoyed its first time around suddenly get this push, seemingly out of nowhere, and come back to life. It was very cool, it was very fun, and I liked how it was making the jump from fiction to graphic novels and also to cosplay. It seemed to express a real zeitgeist of the moment.

SPM: *How have you seen steampunk change over these past years?*

CORY: The big change was from a literary form to a visual form. And that seems to be its major strength. What made steampunk really work wasn't just the playful thought experiment that you could play out in fiction. Fiction's major virtue as a form is interiority. You can hear what people are thinking. That's why so many of the novels that get made into movies end up featuring voiceover narration, because that interiority is the thing that separates fiction from all other forms.

But with steampunk, its major impact isn't knowing how people think when they're thinking in a techno-victorian way, its major impact is seeing what they make in techno-victorian context. So the jump from textual to visual and then to... I don't even know what you call costuming. It's visual but it's live, I guess. That was a very exciting jump for it to have made.

SPM: *You write a lot about maker culture as well. What influence do you think maker culture can have on the world?*

CORY: The motto of *SteamPunk Magazine*, "love the machine, hate the factory," is one of the most provocative statements about technology that I've encountered because it demands that we separate out technology from the way that we use it, the thing that technology *is* from the way that we *use* technology. It says that we can have a factory that is just you. That technology can come about without having to have assembly lines, without having to have bosses, without having to sacrifice all the autonomy that you get as a free agent who isn't part of a larger coordinated system, and still make stuff happen.

Maker culture is exciting today as a kind of response to a couple of things. One is that in a networked society you can solve most problems by finding how someone else has solved a problem nearly identical to your problem and then tweaking it a bit. So we have a kind of just-in-time knowledge world, where you don't really have to know everything, you just have to know how to find things, and when you find them it's very easy to turn them into action.

And then the other thing is that the benefit of automation is that you need fewer labor inputs to accomplish comparable tasks, so there're just fewer jobs around for people to do. The robots have stolen all of our jobs and as a result, a lot of what we do is increasingly non-monetary in nature. And this has been disastrous. It has left a lot of people very poor. But the solution to it can't be make-work jobs, or de-automating the world. I think the solution has to be to increase the extent to which our quality of life is improved by non-financial or non-market activity—stuff that you love doing, not stuff that you're paid to do. And I think making sits in that sweet spot.

SPM: *And what about being a writer? You've never shied from politics, or at least culturally radical ideas, in your writing. What effect do you think creating stories can have on the wider culture?*

CORY: I think writing, and in particular science fiction, puts a lot of meat on the bones of the argument. When we have an argument about technology or an argument about a policy, that argument tends to be very abstract. "I don't want CCTV cameras because they make me feel watched, and when I'm watched I feel less free," is a very abstract thing to say. But if you wait until George Orwell goes along and writes *1984* you can say "I don't like being watched because it's Orwellian." And you get to import into the discussion all of the emotion, all of the feelings, that you experience when you read *1984*. And that makes it possible to bring the argument from these very abstract and difficult to grasp areas into a very concrete and very personal area. And fiction, I think that that is its major impact on debates about policy and technology, and it's one of the reasons I write fiction. I think the main reason I write fiction is that I can't stop, but that's one of the reasons I write fiction.

Séance Through

Thomas Alva Edison is, I'm sure, a man who requires little or no introduction to a large portion of this magazine's readership. In his 84 years of life Edison came up with many, many innovations including the phonograph, the motion picture camera, the stock ticker, the power station, and of course, the light bulb. In fact, Edison is still the fourth most prolific inventor in history, holding 1,093 patents in the USA alone. He also routinely electrocuted numerous animals including an elephant (see *SteamPunk Magazine #2* "To Electrocute an Elephant—How Edison Killed a Century on Coney Island"), but that's another story. All this, as I say, you probably already know. You may even have heard, or read, that Edison's last breath is preserved at the Henry Ford Museum (aka the Edison Institute) in Dearborn, Michigan. Ford having convinced Edison's son Charles to seal a glass vacuum tube of air from the inventor's room shortly after his death in October 1931. Granted, that seems a bit odd, but *memento mori* were not so uncommon then, and Ford was a great friend of Edison's after all. Even so, some might wonder whether it's something Edison senior would have consented to himself. It does seem rather morbid—superstitious even—and surely at odds with the hard scientific logic of a man who was quoted by the *New York Times* in 1910 as saying he had come to the conclusion that *"there is no 'supernatural,' or 'supernormal,' that all there is can be explained along material lines"*.[1] But then again, perhaps not.

The October 1920 issue of *American Magazine* contained an article with the rather attention-grabbing title of "Edison Working on How to Communicate with the Next World" written by one Bertie Charles Forbes (founder of *Forbes Magazine*).[2] In an interview conducted by Forbes and published in *Scientific American* soon after, Edison expressed some rather interesting—some might say surprising—opinions concerning no less a subject than life after death:

> "If our personality survives, then it is strictly logical and scientific to assume that it retains memory, intellect, and other faculties and knowledge that we acquire on Earth [...] I am inclined to

SCIENCE
Edison's Ghost Machine

By John Reppion
Illustration by Michael Barnes (Elblondino)

believe that our personality hereafter will be able to affect matter. If this reasoning be correct, then, if we can evolve an instrument so delicate as to be affected, moved, or manipulated [...] by our personality as it survives in the next life, such an instrument, when made available, ought to record something."[3]

Indeed, it would seem that these were not just idle musings, because in a private journal entry, again dating from 1920, Edison wrote:

"I have been at work for some time building an apparatus to see if it is possible for personalities which have left this earth to communicate with us [...] I am engaged in the construction of one such apparatus now, and I hope to be able to finish it before very many months pass."[4]

A second *Scientific American* piece ran in 1921 in which Edison was quoted as saying:

"I don't claim anything, because I don't know anything [...] for that matter, no human being knows [...] but I do claim that it is possible to construct an apparatus which will be so delicate that if there are personalities in another existence who wish to get in touch with us [...] this apparatus will at least give them a better opportunity."[5]

By this time it seems it must have been pretty widely known that Thomas Alva Edison, one of the greatest inventors of all time, was working on a machine which might prove the existence of spirits or ghosts. The editor of *Scientific American* reportedly received more than 600 letters from readers enquiring about the device.[6] This was big. So, what happened next? The answer is nothing. Absolutely nothing. The machine was never mentioned again during Edison's lifetime and the whole matter seems to have been all but forgotten. Forgotten that is, until old Edison had been in his grave for two years.

Page 34 of the 1933 October edition of *Modern Mechanix* (motto: "Yesterday's Tomorrow Today") bore the intriguing headline "Edison's Own Secret Spirit Experiments." "For thirteen years results of Edison's astounding attempt to penetrate that wall that lies beyond mortality have been withheld from the world, but now the amazing story can be told." The prodigiously illustrated three page piece tells a tale of the "black, howling wintry night in 1920—just such a night when superstitious people would bar their doors and windows against marauding ghosts—"when Edison and a group of scientists and spiritualists "assembled like members of a mystic clan" to test his theories concerning life after death. Edison was, we are told, armed with a powerful lamp whose light was concentrated into a beam and directed at a photo-electric cell which in turn transformed that light into an electric current. Any object passing through that beam of light, no matter how miniscule or insubstantial, would disrupt the electrical current and that fluctuation would be displayed on the dial of a meter connected to the photo-electric cell. This rather disappointingly simple set up was, we are informed, the machine with which Edison sought to prove or disprove the existence of ghosts or spirits once and for all. "When the experiment was ready to begin the spiritualists in the group of witnesses were called upon to summon from eternity the ethereal form of one or two of its inhabitants, and command the spirit to walk across the beam." And the result of this groundbreaking experiment? Well, in the long hours that followed, during which the "wind howled around the corners of the laboratory," the needle, we are told, never so much as wavered. "It was because of these negative results that the news of the amazing experiments was never given out to the world. Edison would not reveal his belief-shattering discoveries to a believing world."[7]

The *Modern Mechanix* piece is written (as you can no doubt tell from the portions quoted above) rather more like a story than a factual article and no author credit is given in the magazine's table of contents. This, coupled with the fact that no members of Edison's "mystic clan" are named, and no sources or references are given, has led some researchers to conclude that the article is, in fact, a piece of fiction (albeit one with a rather anticlimactic ending) woven out of the fragments of information given in interviews and articles published during Edison's lifetime.[8] [9] There is, however, another reason why some who are interested in Edison's paranormal experiments might instantly take the *Modern Mechanix* piece for a fiction: the apparatus described is all wrong.

> Edison was said to be developing a machine that would measure "one hundred trillion life units" in the human body that "may scatter after death."

In a 1921 *New York Times* article Edison was said to be developing a machine that would measure "one hundred trillion life units" in the human body that "may scatter after death."[10] The *Modern Mechanix* article reused much of the material from that earlier piece in explaining what it referred to as Edison's hypothesis of "immortal units." Edison is said to have taken a print from one of his fingers and then to deliberately burn that fingertip so as to remove or alter its print. Later, when the finger had healed, he took a second print which proved to be identical to the original. "From this experiment, Edison got confirmation of his hypothesis that it is these aforementioned 'immortal units' which supervised the re-growth of his finger skin, following out the original design. Man, he believed, is a mosaic of such life units, and it is these entities which determine what we shall be."[11] These "immortal units" then are supposedly what Edison was expecting to break his beam of light (though exactly why Spiritualists would be required to summon them is anyone's guess).

I first began researching Edison's alleged supernatural experiments several years ago. At the time I did quite a bit of Googling around and bookmarked about twenty or so webpages containing relevant information. It was always a subject I meant to come back to but, for one reason or another, I just didn't find the time. A couple of weeks ago, I spotted the folder marked EDISON in my bookmarks and clicked to Open All in Tabs. More than half the links were dead. Of those that remained, most focussed on the idea that Edison could have been working on an apparatus or experiment very much like the one described in the *Modern Mechanix* article—a way of seeing or measuring the hypothesised "life units" or "immortal units." Such an experiment would, naturally, have been doomed to fail. Many of those websites since deleted, and a handful of the links remaining however, focussed on Edison's own journal entry of 1920: "I have been at work for some time building an apparatus to see if it is possible for personalities which have left this earth to communicate with us." Communication with—rather than mere detection of—ghosts or spirits is still believed by some to have been Edison's true goal. *Edison's Spirit Telegraph*, or *Spirit Telephone* are wonderfully evocative terms which still turn up a few interesting search results. "Thomas Edison was trying to build a machine to talk to the dead," writes one blogger, "I can recall first coming across those very words in an old, dusty book back in the 1970s."[12] "After his death, the plans for the apparatus could not be located. Many have searched extensively for the components, the prototype or even the plans to the machine but have never found them," concludes another.[13]

Some, however, have expressed doubts as to the authenticity of the 1920 diary entry, much of which seems like a mere reproduction of portions of the original *Scientific American* interview with the "I have been at work for some time building an apparatus [...]" paragraph tacked on at the end.[14] Furthermore, there is one very important piece of evidence which many seem to have overlooked, whether accidentally or wilfully. In an interview published in the *New York Times* in 1926 Edison was asked about the comments he'd made six years earlier concerning the prospect of investigating the survival of spirits after death, to which he replied "I really had nothing to tell him [Forbes], but I hated to disappoint him so I thought up this story about communicating with spirits, but it was all a joke."[15] And so, like the *Modern Mechanix* piece before it, our own tale of Edison's Secret Experiments ends with something of anticlimax—the whole thing was merely a hoax. But then again, perhaps not.

In 1941, a séance was supposedly conducted in New York in which a spirit claiming to be that of Thomas Alva Edison made itself known. This spirit, it is alleged, named certain associates (members of the "mystic clan," if you will) who apparently still had in their possession the missing plans for, and elements of, his machine. These people were located. A prototype was built. It did not work. This prototype somehow passed into the possession of one J. Gilbert Wright—a General Electric researcher whose claim to fame was the discovery/invention of a special kind of silicone putty. Wright, it is said, spent the rest of his life trying to perfect the machine. In some versions of the tale, Wright frequently consults Edison's ghost, via regular séances to get his advice on how the machine might be improved. When Wright finally passed on in 1959 all trace of the machine is said to have vanished.[16, 17, 18, 19, etc.] A more fittingly farfetched end to a tall tale? Perhaps. Even so, surely I'm not the only one left wondering; what if there was just one component needed to complete the machine that Wright could never lay his hands on? Something another member of Edison's "mystic clan" wasn't willing to part with. Something, perhaps, as small and innocuous seeming as one very specific glass vacuum tube. ✺

Endnotes

1. HTTP://WWW.PARANORMAL-ENCYCLOPEDIA.COM/E/THOMAS-EDISON/
2. HTTP://WEB.ARCHIVE.ORG/WEB/20100328084800/HTTP://WWW.MUSEUMOFHOAXES.COM/HOAX/HOAXIPEDIA/THOMAS_EDISON_AND_HIS_SPIRIT_PHONE/
3. HTTP://WWW.PARANORMAL-ENCYCLOPEDIA.COM/E/THOMAS-EDISON/
4. DAGOBERT D. RUNES (EDITOR): THE DIARY AND SUNDRY OBSERVATIONS OF THOMAS A. EDISON (NEW YORK PHILOSOPHICAL LIBRARY, 1948)
5. HTTP://LONGSTREET.TYPEPAD.COM/THESCIENCEBOOKSTORE/2011/02/EDISON-AS-GHOSTBUSTER.HTML
6. HTTP://WWW.GEREPORTS.COM/EDISONS-FORGOTTEN-INVENTION-A-PHONE-THAT-CALLS-THE-DEAD/
7. HTTP://BLOG.MODERNMECHANIX.COM/2006/08/14/EDISONS-OWN-SECRET-SPIRIT-EXPERIMENTS/
8. HTTP://STRANGEANDSPOOKYWORLD.WORDPRESS.COM/2011/07/16/THOMAS-EDISON-GHOSTBUSTER/
9. HTTP://WWW.PARANORMAL-ENCYCLOPEDIA.COM/E/THOMAS-EDISON/
10. HTTP://WWW.PARANORMAL-ENCYCLOPEDIA.COM/E/THOMAS-EDISON/
11. HTTP://BLOG.MODERNMECHANIX.COM/2006/08/14/EDISONS-OWN-SECRET-SPIRIT-EXPERIMENTS/
12. HTTP://STRANGEANDSPOOKYWORLD.WORDPRESS.COM/2011/07/16/THOMAS-EDISON-GHOSTBUSTER/
13. HTTP://WWW.PRAIRIEGHOSTS.COM/OH-MILAN.HTML
14. HTTP://WWW.PARANORMAL-ENCYCLOPEDIA.COM/E/THOMAS-EDISON/
15. HTTP://WWW.PARANORMAL-ENCYCLOPEDIA.COM/E/THOMAS-EDISON/
16. HTTP://WEB.ARCHIVE.ORG/WEB/20090523133345/HTTP://THEBEYOND.INFO/EDISON.HTM
17. HTTP://WWW.GHOSTEYES.COM/THOMAS-EDISON-PARANORMAL
18. HTTP://WWW.GHOSTVILLAGE.COM/GHOSTCOMMUNITY/INDEX.PHP?SHOWTOPIC=19802
19. HTTP://UFOEXPERIENCES.BLOGSPOT.COM/2007/04/SURVIVAL-OF-SPIRIT-PERIPHERAL.HTML

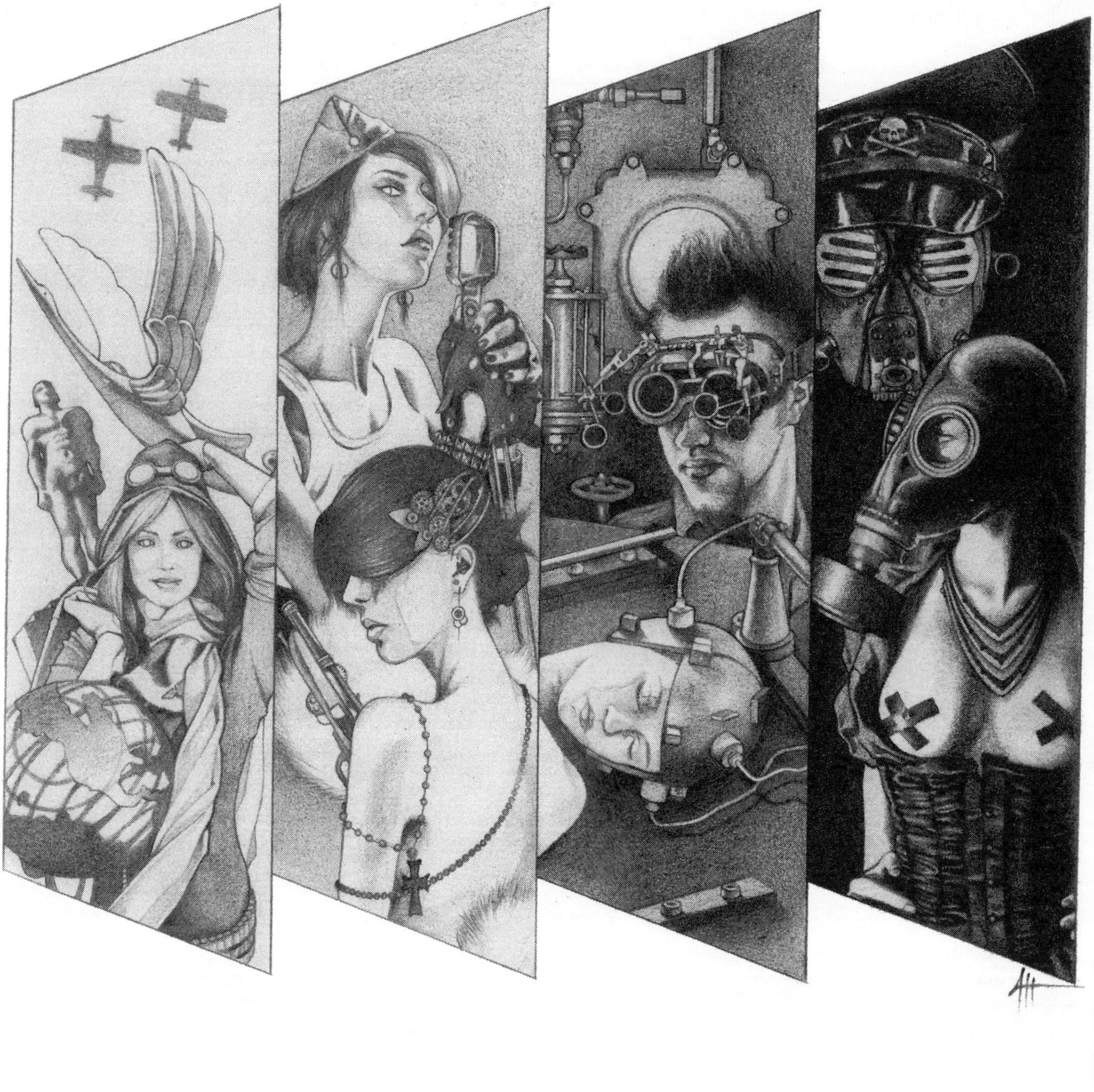

a Dieselpunk Cookbook

By Larry Amyett, Jr.
Illustration by Allison M. Healy

Those who read my article, "A Dieselpunk Primer," in issue #8 of *SteamPunk Magazine* already have a basic understanding of dieselpunk. Dieselpunk essentially consists of three components: First, dieselpunk contains "diesel," which refers to the history and aesthetics of 1920s through the early 1950s, which dieselpunks refer to as the Diesel Era. Second, dieselpunk is contemporary in origin—the Diesel Era is providing the inspiration and source material for the process of creating something new and original. And thirdly, there's the "punk," which allows for genre-twisting and -mixing along with providing an emphasis on independence and going outside the mainstream.

As I looked back on my previous article, I realized that it lacked an in-depth explanation of the various manifestations, or what has been called flavors, of dieselpunk.

A consensus in the dieselpunk community has developed that there are generally two major flavors, or styles, of dieselpunk: Ottensian and Piecraftian. These names originated from the work of dieselpunk pioneers Nick Ottens and Bernardo Sena who wrote the article "Discovering Dieselpunk" in the first issue of the now extinct online magazine *The Gatehouse Gazette*. (At the time, Mr. Sena was writing under the pseudonym "Mr. Piecraft.")

The best explanation of the difference between Ottensian and Piecraftian dieselpunk is the popular saying of looking at a glass as being either half-empty or half-full. While Ottensian dieselpunk varies from positive or dark, it ultimately has a sense of hope and success to it. By contrast, Piecraftian dieselpunk generally has a pessimistic and apocalyptic aspect to it. Each of these flavors are divided into various sub-classes.

I think it's important to point out that the term "flavor," which Ottens and Sena used in their *Gatehouse Gazette* article, is very appropriate for this topic. Just as a recipe includes a variety of flavors from many ingredients, most dieselpunks mix the different flavors to suit their personal tastes. Hence, this article aims to provide the reader with what I call a cookbook of the basic flavors that they can use for creating their individual recipe for dieselpunk.

World of Tomorrow: Hopeful Ottensian Dieselpunk

"Believe in life! Always human beings will live and progress to greater, broader and fuller life."
—W. E. B. Du Bois

Without a doubt, the Diesel Era saw amazing advances in technology and science. The first commercial radio station began broadcasting in 1920, automobiles became commonplace as their sophistication increased and costs dropped, and the Great War had turned the primitive planes of the Wright Brothers into impressive flying machines. It was also an era of great scientific minds such as Einstein and Heisenberg.

In the area of the arts, the Diesel Era saw amazing developments. It's called the Golden Era of motion pictures and for a good reason, with actors such as Charlie Chaplin, Katherine Hepburn, James Cagney, and Ingrid Bergman on the screen. The Diesel Era saw the growth of jazz and the blues as well as the appearance of big band and swing. It was the age of dada, surrealism, and art deco.

For many people, especially during the 1920s, the possibilities of human progress seemed limitless. It's on this positive and optimistic vision of the world, a view of humanity progressing to greater heights thanks to human knowledge and ingenuity, that Hopeful Ottensian dieselpunk centers.

Without a doubt, the best example of Hopeful Ottensian dieselpunk in cinema is the movie *Sky Captain and the World of Tomorrow*. According to Kerry Conran in an article at FilmFreaks.net, "the title refers to the World Expo and the spirit of that was looking at the future with a sense of optimism and a sense of the whimsical, you know, something that we've lost a lot in our fantasies. We're more cynical, more practical, which they lacked. They never thought about how practical it was to dock a zeppelin at the Empire State Building, but they did it anyway because it was the future. In that regard I think what this film attempts to do is to take that enthusiasm and innocence and celebrate it—to not get mired in the practicality that we're fixated upon today."

Sky Captain captures the optimism of Hopeful Ottensian dieselpunk with Sullivan's modified Curtiss P-40, Dex's ray gun, and the armada of British flying airstrips. Even with Totenkoff's use of technology for evil, there's an underlying positive theme to the movie: even one villain using technology for evil couldn't hold back human progress.

In Hopeful Ottensian dieselpunk, hard times are just bumps on the road of progress. The future is positive and humanity holds the key to its own salvation.

> In Hopeful Ottensian dieselpunk, hard times are just bumps on the road of progress. The future is positive and humanity holds the key to its own salvation.

Innocence Lost: Dark Ottensian Dieselpunk

"All the time the flapper is laughin' and dancin', there's a feelin' of tragedy underneath..."
—Clara Bow

Modernity, with its freedoms and challenges, obviously didn't originate in the Diesel Era. But the same advances that the Hopeful Ottensians celebrate helped contribute to a crisis in which America had to face modernity in ways that it never had to before. America's response to modernity during the Diesel Era was what we might today refer to as a culture war. In the forward to Loren Baritz's book *The Culture of the Twenties*, Alfred F Young and Leonard W. Levy wrote that the characteristics of the 1920s were "products of a pervasive clash between the values of a new urban, urbane, and modern civilization, and the pieties of small town, provincial America."

Though they were fewer in numbers, American political power in the twenties rested with small town conservatives who fought ferociously against modernity. As a result, they were able to force through their agendas such as Prohibition and the "criminal syndicalism" laws that resulted in events such as the Palmer Raids. In addition, the twenties saw a massive growth in American Nationalism and membership in the Ku Klux Klan.

The American cosmopolitan intellectuals, writers, and artists, who were more receptive to modernity, responded with revulsion to this right-wing political activity, pulling away rather than pushing back. Many of those who could afford it fled to Europe, while others dug in and created pockets of progressive and artistic communities within the major cities.

Caught in between these two extremes were the majority of Americans. According to Baritz, "The middle term was the world of the Jazz Age, flapper, speakeasy, and the rest. Reaching both forward and backward, it knew it was not truly of either world. It was the booming New Era, the Roaring Twenties. But it too was caught by the power of the village; it had to consume its booze secretly lest the village law cause embarrassment."

The 1930s saw the world plunge into one of the worst economic crises of modern history, the Great Depression, and Mother Nature herself seemed to declare war on the American people. For years, farmers had overplanted without rotating the crops and had removed the native grasses. This, along with a period of extended drought, resulted in the nightmare that was the dust bowl.

The international political scene was just as bleak. Stalin was terrorizing the people of the Soviet Union. In Italy Mussolini claimed to make the trains run on time but ruled the nation with an iron fist [*and didn't make the trains run on time, either. —ed*]. The Nazis rose to power in Germany, terrorizing the world with their anti-Semitism, hatred, and military expansionism. The Second Spanish Republic fell to the fascists in the Spanish Civil War. In Japan the military gained power through "government by assassination" and began expanding its "Greater East Asia Co-Prosperity Sphere," which, despite its propagandistic name, was nothing more than military conquest.

By the 1940s, the Second World War shattered any hope of a positive future as evil incarnate threatened to dominate the entire globe. Never in human history had there been death and destruction on such a scale. Millions of lives were lost, whole cities were obliterated, and weapons were created of such magnitude that few had dreamed possible.

It's in these dark and violent times that Dark Ottensian dieselpunk inhabits.

I share the opinion of Ottens and Sena that the 1994 motion picture *The Shadow* is an excellent cinematic representation of Dark Ottensian dieselpunk.

In the opening scene of *The Shadow*, we see Lamont Cranston, played by Alec Baldwin, living in Tibet as a cruel and heartless opium drug lord. After being kidnapped by the Tulku, a local holy man, he's told by the Tulku that he knows that Cranston's been tormented by his black heart and was always in great pain. We then hear the famous line, in this case spoken by the Tulku, that Cranston knows "what evil lurks in the hearts of men" because Cranston had "seen that evil in [his] own heart." Cranston was destined for redemption and goes back to America to continue the eternal struggle against evil.

Throughout the movie, Cranston is a tormented man. When Margot Lane (Penelope Ann Miller) tells him, "I'm not afraid of you" Cranston replies, "But I am." Later, he tells Margo, "You have any idea what it's like to have done things you can never forgive yourself for?"

The world of *The Shadow* is one of drug lords and mobsters. Crime is rampant and the inept police force is incapable of protecting its citizens. With society filled with violence and corruption, the people's only hope is a vigilante armed with ancient knowledge and occult powers.

Unlike in Hopeful Ottensian, in Dark Ottensian neither human technological progress nor modernity gives us hope for a better future. Instead, this drive for domination and control

> in Dark Ottensian neither human technological progress nor modernity gives us hope for a better future. Instead, this drive for domination and control over both nature and ourselves is setting us up for destruction. The greater we strive to exert control and establish order, the more chaos is born.

over both nature and ourselves is setting us up for destruction. The greater we strive to exert control and establish order, the more chaos is born.

At the end of the movie, Cranston defeats Khan by learning to master the one thing that he never could before: himself. However, he achieves this mastering not by exerting greater control, but instead by letting go. He lets go of his past and the demons that haunt him and stops trying to control the Phurba by brute force.

While the world of Dark Ottensian isn't positive, it's not hopeless either. Hope for humanity in Dark Ottensian lies not in modernity's drive for control and technological progress, but by learning to let go of this desire for control and the demons that hold us back.

Tyranny of the Mind: Dystopian Piecraftian Dieselpunk

"Our landings have failed and I have withdrawn the troops. My decision to attack at this time and place was based on the best information available. The troops, the air and the Navy did all that bravery could do. If any blame or fault attaches to the attempt it is mine alone."

—*A draft speech written by General Dwight D. Eisenhower just before Operation Overlord, in case the operation failed*

History is full of turning points. With a wrong military decision or simple bad luck, a democratic nation might have lost a war that in reality it had won or a democracy might have failed where one had flourished. Such possibilities have intrigued writers and historians for years.

Thankfully, the Allied invasion of Normandy in World War II was a success. Yet, some writers have postulated on the "What if?" scenario of a possible failure of the Allies in World War II. For example, Peter G. Tsouras edited *Third Reich Victorious* in which various historians wrote short stories presenting scenarios in which Germany could have won World War II.

There are other possible Diesel Era nightmare scenarios. In 1935, Sinclair Lewis wrote the classic novel *It Can't Happen Here* in which a character named Buzz Windrip was elected president of the United States and turns America into a dictatorship. The author loosely based the character of Windrip on the real life Huey Long, who challenged FDR for the Democratic nomination but was assassinated. Nor can we forget the most famous Diesel Era dystopia novel of all time: *1984* by George Orwell.

> Thomas Jefferson wrote, "I have sworn upon the altar of God, eternal hostility against every form of tyranny over the mind of man." While Jefferson swore to fight against such tyranny, Dystopian Piecraftian dieselpunk uses the genre to take an unflinching look at its very real possibility.

The Dystopian Piecraftian sub-class of dieselpunk takes such possibilities seriously. This sub-class of dieselpunk could best be defined as a contemporary mixture of Diesel Era aesthetics with either an alternate history or future that's often centered on, to quote Mr. Sena, "a world in which the enemy or ruling authoritarian state are a controlling force, unveiling a truly hopeless dystopian future."

At the start, it's immediately obvious that Terry Gilliam's Dystopian Piecraftian dieselpunk movie *Brazil* is going to be unique. The style of dress is distinctly Diesel Era 1940s, full of fedoras and suits. While the society has computers and robotics, the technology is a bizarre mix of realistic machines patched together with unrelated parts that (barely) function along with some wonderfully odd machines and vehicles, all of which seemed to have come straight from the Diesel Era magazine *Modern Mechanics*.

The politics of the country is a strange Orwellian-style totalitarian society in which, while apparently capitalistic, the State bureaucracy monopolizes everything with "Central Services" and controls all knowledge with the Ministry of Information. Part of the Ministry of Records is the division of "Information Retrieval," which is a Gestapo style organization that uses repeated terrorist attacks as an excuse to torture and oppress the people.

Thomas Jefferson wrote, "I have sworn upon the altar of God, eternal hostility against every form of tyranny over the mind of man." While Jefferson swore to fight against such tyranny, Dystopian Piecraftian dieselpunk uses the genre to take an unflinching look at its very real possibility.

Becoming Death: Post-Apocalyptic Piecraftian Dieselpunk

"Now I am become Death, the destroyer of worlds."
—*Julius Robert Oppenheimer quoting the Bhagavad Gita after the first atomic bomb was detonated in 1945*

As if the possibility of world tyranny wasn't bad enough, the Diesel Era brought us the possibility of something even more ominous. We learned that humanity not only had the will but now the power to destroy the world.

By 1945, the world had seen the worse that humanity could become. The two World Wars, the Holocaust, and other atrocities had shown the depths of depravity to which the human race could sink. However, as though these horrors were not enough, the end of World War II saw that the human species had, with the invention of nuclear weaponry, finally gained the ability to annihilate all life on earth. This ability to destroy, when combined with the evil humanity had shown it was capable of in prior wars, warned of the potential for a disaster of an unimaginable magnitude.

In the prologue of Ralph Bakshi's Post-Apocalyptic Piecraftian dieselpunk animated masterpiece *Wizards*, we learn that five terrorists had exploded a nuclear device, triggering a chain of nuclear attacks that had destroyed the world and made it nearly uninhabitable. The only inhabitants left were mutants created by centuries of radiation. Then, after two million years, the elves and fairies returned to occupy the healthy green lands while the mutants continued to live in the radioactive wastelands.

One night the fairy queen gives birth to two wizard babies: a wise and good wizard named Avatar and an evil wizard named Blackwolf. After their mother's death, the two wizards, now adults, battle. Avatar wins the fight and drives Blackwolf into the wasteland. For over five thousand years Blackwolf studies black magic and digs up pre-holocaust technology in the hopes of finding the means of conquest, but his discoveries are not enough to motivate his army of mutants to defeat the powers of magic. Then Blackwolf discovers something special.

Until this point in the film, after being told by the narrator that the story concerns the ongoing battle between technology and magic, the animated movie seems more renpunk than dieselpunk. The first hints of dieselpunk appear when we see a swastika painted under Blackwolf's throne. At one point Blackwolf pulls back a curtain consisting of a Nazi flag and announces, "Attention, leaders of tomorrow's master race!" He then uses a projector to show Nazi Germany propaganda along with German World War II military assaults, which we learn is a dream machine that paralyzes the elves and fairies with fear to be slaughtered by his army of mutants and demons.

In Post-Apocalyptic Piecraftian dieselpunk, we find a focus on the ultimate failure of humanity: its own destruction. It is the ominous vision of a dead or dying world combined with a Diesel Era aesthetic or element that is the Post-Apocalyptic Piecraftian sub-class of dieselpunk.

Conclusion: Dieselpunk as a Message for Our Times

As stated, Hopeful Ottensian dieselpunk looks at human progress from a positive standpoint. While human technological progress may advance at different paces, and at times might even slide backwards, the general direction is forwards. Many people, including myself, owe our lives to the miracles of modern medicine. Hopeful Ottensian serves an important purpose of reminding us that while the world is far from perfect, there are many positive aspects to our current times and that we need to avoid romanticizing the past. That there never were any "good old days."

Where the Hopeful Ottensian is positive about human progress, Dark Ottensian is a cold, hard, slap of reality. The amazing progress that Hopeful Ottensians point to is unevenly distributed—there are a vast number of people in the world whose lives are, to borrow a phrase from Thomas Hobbes, "poor, nasty, brutish, and short." There's widespread crime, crushing poverty, war, genocide, disease, and suffering. In fact, the very advances in technology and increases in productive ability due to industrialization and modernity have toxic side effects such as pollution and the oppression of workers. Just like the 1920s, we have our own culture wars and the recent economic collapse is reminiscent of the collapse of the 1930s. Then when we try to fix the problems it all too often turns out as Robert Burn wrote: "The best laid schemes of mice and men / Go oft awry." If Hopeful Ottensian keeps us from becoming luddites, Dark Ottensian is a warning against hubris.

The warning in Dystopian Piecraftian dieselpunk is that freedom and human rights are fragile. As our world has become smaller and technology has tied us closer together, the potential for the loss of individual freedom has grown considerably. There has been a growing fear of how much freedom we should sacrifice for safety. The Left fears governmental and corporate intrusion into our lives while the Right fears bureaucratic control in our economic activities. Dystopian Piecraftian dieselpunk reminds us that a free people must always be vigilant against tyranny.

While the ability to destroy the world through nuclear weaponry became a reality during the Diesel Era, we've since come up with new ways of imagining armageddon. Besides nuclear war, we now worry about global warming, overpopulation overcoming the world's resources, and a worldwide pandemic created by increased global travel or genetic experimentation, along with other possible nightmare scenarios. Post-Apocalyptic Piecraftian fiction reminds us to cherish our time on earth for life is fragile and we have no idea what the future might hold.

Dieselpunk is more than entertainment or a lifestyle. We can use the different flavors from the dieselpunk cookbook to explore real world issues that are relevant to our world today.

THE SHAKER OF BONES

AN INTERVIEW WITH CHERIE PRIEST

By Margaret Killjoy
Illustration by Sarah Dungan

Cherie Priest is the author of a dozen novels, including the Hugo- and Nebula-nominated and Locus-winning steampunk novel Boneshaker and three other books in her Clockwork Century universe. She's one of the smartest and most well-spoken steampunks around, what's more, and seems to have a reputation for saying things worth hearing and worth repeating. So I was quite happy when she took the time to answer a couple of questions for our readers.

STEAMPUNK MAGAZINE: So you're in many ways steampunk's runaway successful novelist. But success didn't just find you overnight, you've really put in your time. How many novels was it you've written before Boneshaker? Can you tell me the story of that?

CHERIE PRIEST: Boneshaker was my seventh published book—and it was never really contracted. It's kind of a long story, but to make it short: I was working on my fifth and last book under contract for Tor—a book which was originally an altogether different story—and I started noodling on Boneshaker on the side. My editor saw me talking about this new project online, and asked to take a look at it. She liked what she saw better than what was already in progress, so she suggested that we switch out the books.

At the time, I was half afraid that it'd be the last book I ever had published under my own name; my previous titles hadn't sold well at all, and I was in a bit of a death spiral from a career standpoint. But then... well, then it sort of took off. No one was more surprised than me! But I'm tremendously grateful for how everything panned out.

SPM: I actually kind of blame you for the recent trend of American Civil War era steampunk, but of course you do it masterfully. What made you decide to set your work over here instead of England?

PRIEST: I felt like everyone else was doing steampunk set in England, and there seemed to be this broad perception that if it wasn't English, it didn't count. I thought it'd be fun to challenge that—and this sentiment collided with my move to Seattle (more or less). Seattle has some crazy local history, exactly the kind of thing that was ideal to mine for fun details. So a general inkling that "American steampunk might be nice" met up with "Seattle sure was weird during the 19th century," and the rest was... wait, if I say "history" that'll be corny, right?

Anyway, by restructuring the war—and subsequently, western expansion and the entirety of the frontier—I gave myself a whole new universe to play in.

SPM: Boneshaker is optioned and might become a hollywood movie? What's that like for you?

PRIEST: Allegedly the movie folks are going into production before terribly long, maybe next year or so. They have a script (by John Shepherd, writer of HBO's Nurse Jackie), and they have the financing nailed down... now it's just a waiting game. Obviously I'm thrilled silly—but it's also somewhat stressful, because there's so much uncertainty. People have so many questions about the movie project, and I have so few answers.

SPM: For years, you were the blue-haired author. What was it like to have that be an identity lofted upon you? How has the way the world interacts with you changed since you've moved to a more understated aesthetic?

PRIEST: I enjoyed the blue hair quite a lot, and would probably still do it if it hadn't gotten so expensive and time-consuming. I stained a hundred pillowcases, and ruined a few scarves and shirts just by walking in the rain. And besides, it was terrible for my hair—which began to dry out and break off. So, alas. It had to go.

In all honesty, it was terribly convenient—I traveled a lot, visiting places where I didn't know anyone; so I'd just say that I had blue hair and red glasses... and everyone knew how to pick me out of a crowd. (Except for once, at an event in San Francisco. I was one of four women matching roughly that same description.) These days I'm back to ordinary brunette, which is definitely easier on my wallet and bedding, but I do miss it sometimes. Sometimes I feel positively invisible... but the older I get, and the more confident I become as a storyteller, the more comfortable with that I become.

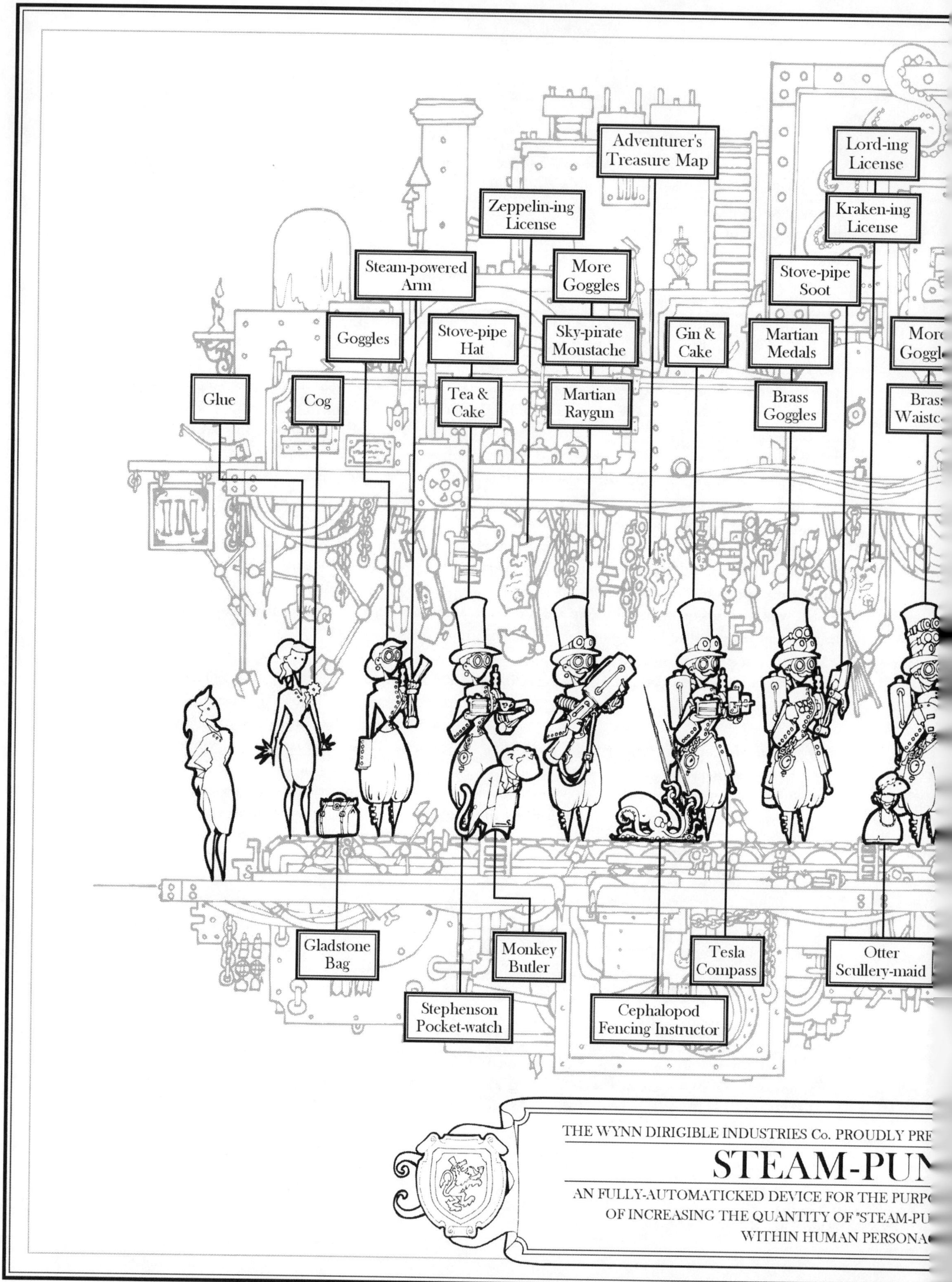
Adventurer's Treasure Map
Lord-ing License
Kraken-ing License
Zeppelin-ing License
Steam-powered Arm
More Goggles
Stove-pipe Soot
Goggles
Stove-pipe Hat
Sky-pirate Moustache
Gin & Cake
Martian Medals
More Goggl
Glue
Cog
Tea & Cake
Martian Raygun
Brass Goggles
Brass Waistc
IN
Gladstone Bag
Monkey Butler
Tesla Compass
Otter Scullery-maid
Stephenson Pocket-watch
Cephalopod Fencing Instructor
THE WYNN DIRIGIBLE INDUSTRIES Co. PROUDLY PRE
STEAM-PUN
AN FULLY-AUTOMATICKED DEVICE FOR THE PURPO
OF INCREASING THE QUANTITY OF "STEAM-PU
WITHIN HUMAN PERSONA

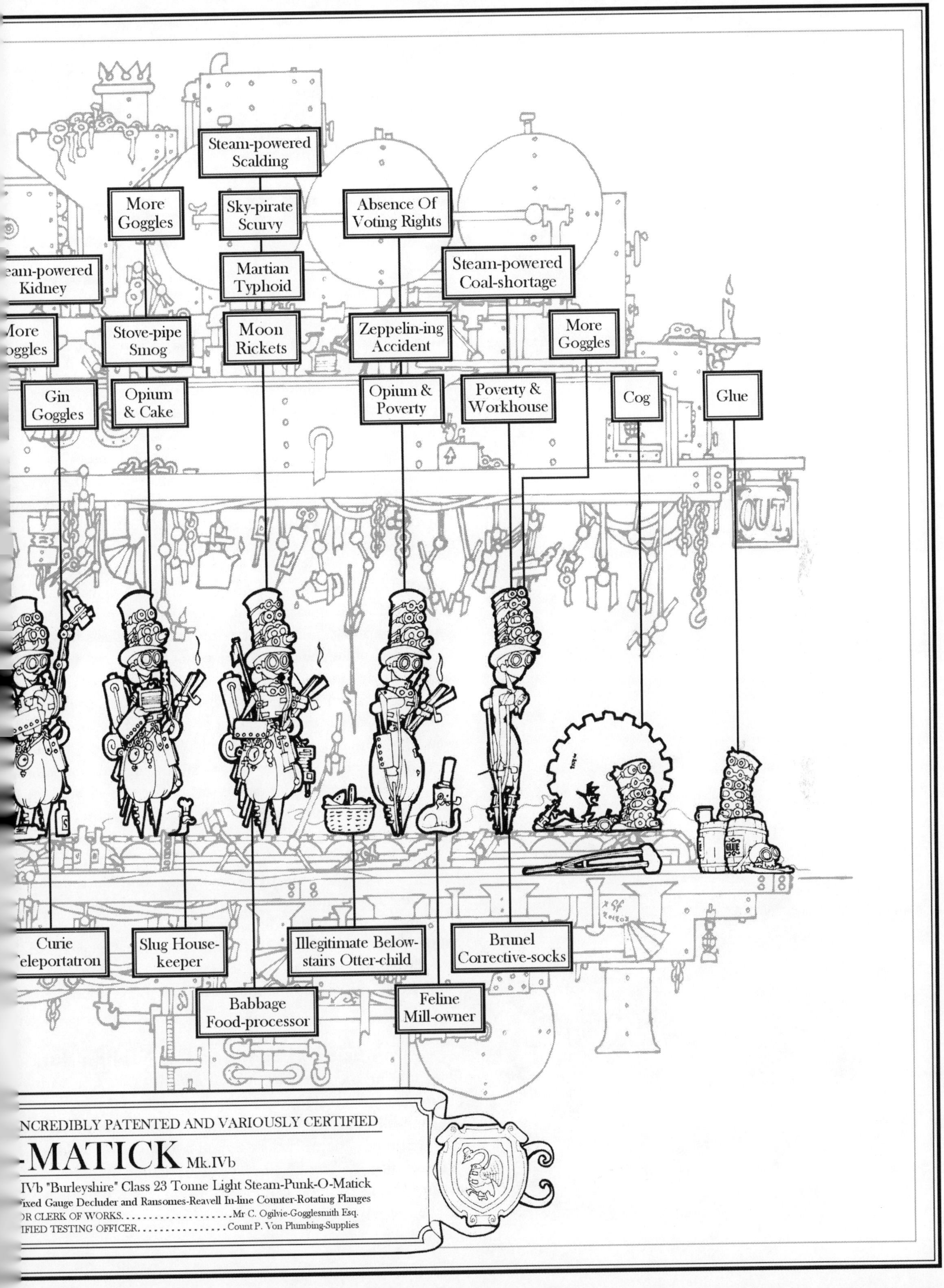

Steam-powered Scalding
More Goggles
Sky-pirate Scurvy
Absence Of Voting Rights
eam-powered Kidney
Martian Typhoid
Steam-powered Coal-shortage
lore oggles
Stove-pipe Smog
Moon Rickets
Zeppelin-ing Accident
More Goggles
Gin Goggles
Opium & Cake
Opium & Poverty
Poverty & Workhouse
Cog
Glue
OUT
GLUE
Curie eleportatron
Slug House-keeper
Illegitimate Below-stairs Otter-child
Brunel Corrective-socks
Babbage Food-processor
Feline Mill-owner
NCREDIBLY PATENTED AND VARIOUSLY CERTIFIED
-MATICK Mk.IVb
IVb "Burleyshire" Class 23 Tonne Light Steam-Punk-O-Matick
ixed Gauge Decluder and Ransomes-Reavell In-line Counter-Rotating Flanges
OR CLERK OF WORKS. . . . . .Mr C. Ogilvie-Gogglesmith Esq.
IFIED TESTING OFFICER. . . . . .Count P. Von Plumbing-Supplies

CREATIVE

By Douglas Summers-Stay
Illustration by Tina Back

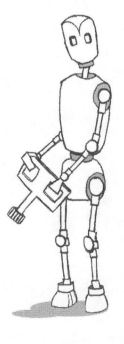

WHAT WOULD IT TAKE TO MAKE A MACHINE THAT WAS CREative? Before the industrial revolution began, the question would have been difficult to ask, since the term "creative" as a capacity of the artist only began to be used at that time. It's still a difficult question to answer, tied up in our ideas of art, and novelty, and the implications of the mechanical and the spiritual. In their attempt to automate the role of the artist, inventors during that time were the first to confront some questions that have only grown in importance as technology becomes more capable.

This article describes a handful of inventions from the 17th through the 19th century that could generate original poetry, music, and mathematical proofs. They have been largely forgotten by history, falling as they do into the cracks between art and technology, between science and magic. To be sure, machines that could be considered as falling into this class were built long before this time; many divination devices composed a fortune based on recombination of terms, for example, and even wind chimes can be considered a type of aleatory music. However, during this time period the advancement in mechanical fabrication techniques, and a social fascination with the potential of machinery led to a variety of fascinating devices, often unique prototypes built by an inventor working alone.

With the invention of the computer, such inventions have become very easy to create in software. And in my work as a robotics research scientist, I see very rapid progress being made in enabling machines to respond to a world they are increasingly able to perceive. Yet many of the same questions remain that were first brought up as society responded to machines that duplicated some of the functions of the creative mind.

Athanasius Kircher and the Aeolian Harp

One of the first devices that could create music (of a sort) without human intervention was the Aeolian harp, an invention by Athanasius Kircher. Kircher was a Jesuit scholar who wrote on an astonishing variety of scientific and religious subjects in the 17th century. The books are eclectic, intriguing, occasionally occult. Perhaps the most comparable figure who is well known to the general public today would be Leonardo DaVinci.

MACHINES

Kircher's books were enormously popular and respected in his day, but as the first real experimental scientists (such as the Royal Society in England) began to test his pronouncements, they found he was wrong as often as right. His interpretations of hieroglyphics, for instance, are comically bad. His reputation was damaged and he was largely forgotten. But he shows up again and again in the history of artificial creativity. Through him we know of many of the ancient automata, methods of automatic composition, the first idea of the kaleidoscope, systems for automatic translation, and many other areas of applied technology that didn't seem worthy of study by later (more theoretical and less whimsical) scientists.

Kircher belonged to the tradition of natural magic, which overlapped more or less with science during this time. Perhaps the closest thing to natural magic today would be a science museum show, or movie special effects: the goal was to astound and to amaze by the use of hidden devices, in a way that led the viewers to curiosity about the explanation. This was in contrast to natural philosophy, which descended from the philosophical tradition which had no place for experiment and was more concerned with geometric and logical demonstrations. We tend to think of natural philosophy gradually becoming science as it began to make use of experiment, but it was only by adopting the methods, technology, and tools invented for natural magic that it was able to do so. Newton's experiments with prisms, for example, were only possible because he was able to buy the prisms which were already being used for natural magic.

Kircher called the Aeolian harp "Machinamentum X" and "Machina Harmonicam Automatam." It consists of several strings stretched over a sounding board. The harp is placed where wind can blow over it, often in front of a window that has been left open a crack. The eddies introduced into the wind by the first string cause the others to vibrate. The name "Aeolian" comes from the story of king Aeolus in the Aeneid, who trapped the winds in the caves of a mountain and forced them to do his bidding.

The Aeolian harp became popular in England and Germany with the rise of the Romantic Movement in the late 18th century. It has been called "The Romantics' scientific instrument." The Romantics saw the wind harp as a symbol of creative inspiration by nature. For example, it is featured in "Ode to the West Wind" by Shelley, and "Dejection: an Ode," and "The Aeolian Harp," by Coleridge. Coleridge made explicit the analogy between mental operations and the blowing of the wind:

> And many idle flitting phantasies,
> Traverse my indolent and passive brain,
> As wild and various, as the random gales
> That swell and flutter on this subject Lute!
> And what if all of animated nature
> Be but organic Harps diversly fram'd
> That tremble into thought as o'er them sweeps
> Plastic and vast, one intellectual breeze,
> At once the soul of each and God of all?

The sound of the wind harp is something ghostly and alien, and has a tendency to make the hair on the back of your neck stand on end. Coleridge called it "a soft floating witchery of sound." They were associated with inspiration by ghosts and spirits. Aeolian harps illustrate artificially creative machines' greatest strength and greatest weakness—the property of being literally inhuman, inconceivable within the usual ideas of what art should be.

The Componium

While the Aeolian harp made sequences of eerie tones, it followed its own natural rules, and bore little resemblance to traditional melodic and harmonic structures. Building a device that could compose under this stricter and more human system would take an engineering genius who also had a deep understanding of music.

Diedrich Winkel was a solitary organ-maker and inventor, born in 1777 in Germany. He never married, but had a successful business in Amsterdam building barrel organs. Winkel led a quiet life, and though he had a reputation as something of a genius among other builders, was little known outside those circles.

In contrast, another inventor of musical devices, Johann Maelzel, was a showman and very famous, traveling all over

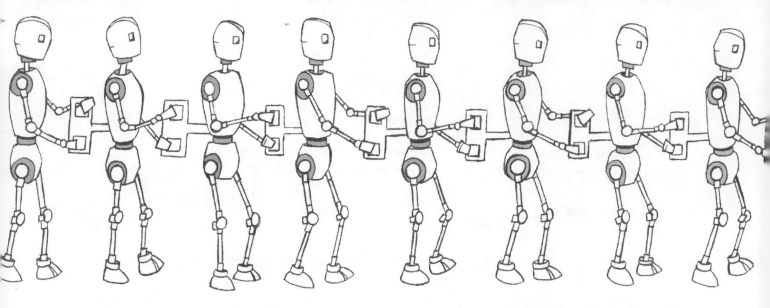

Europe and plastering the alleys with advertisements to see his odd collection of mechanical wonders. He was spoken highly of by Salieri and Beethoven for some of his musical inventions.

When Maelzel was touring Amsterdam, Winkel received a visit from him. Winkel was apparently excited to have the chance to show off his inventions to an appreciative audience, and showed Maelzel a simple but useful tool he had invented: the world's first effective metronome, the pyramidal model still often seen today. Maelzel had himself invented a metronome, but it was rather impractical, and Maelzel knew it.

Maelzel turned out to be something of a scoundrel, and began selling metronomes on Winkel's design, claiming them as his own invention. Despite a lawsuit, Winkel was never able to stop Maelzel from selling these, and to this day the design, still popular, is referred to as Maelzel's metronome.

The invention Maelzel seems to have been most proud of, however, was his Orchestrion. The Orchestrion was a kind of band-in-a box. It had a dozen instruments, all driven by a changeable barrel, like a barrel organ—a cylinder with pins sticking out that as it revolves, triggers notes to be played. Maelzel got Beethoven to compose a piece for his Orchestrion, and, true to character, failed to pay him for it.

(Maelzel also purchased for his traveling show an invention that should be familiar to anyone interested in steampunk—the famous Mechanical Turk. The story of the chess-playing Turk is wonderful, involving everyone from Catherine the Great to Edgar Allen Poe; but the point here is that it is yet another example of Maelzel's propensity to fraud, since instead of the automaton he claimed, it was a puppet controlled by a chessmaster hiding in the cabinet beneath the board.)

Winkel apparently decided that the best way to get back at Maelzel would be to show once and for all that he was a greater inventor than Maelzel. He too built a device similar to the Orchestrion, but with a twist—after playing through a piece once, it would begin improvising on the melody, each time playing a unique variation!

> He too built a device similar to the Orchestrion, but with a twist—after playing through a piece once, it would begin improvising on the melody, each time playing a unique variation!

This was accomplished by the following system: for each two measure section of the piece, eight different possible variations appeared on the barrel. As part of the gearing, a disk was allowed to spin freely for a second before a leather band tightened back down on it, and depending on where this disk was pointing at the end of the spin, a different variation on the section would be selected from the barrel. There were two barrels, and each played its two measure section as the other was selecting the next section to be played.

The Componium, as it was called, was something of a revelation to some of the people who paid to see it play. For example, the French education reformer Jean-Joseph Jacotot wrote about the Componium as a metaphor for the subconscious mind. In his book on music, he describes the Componium (representing genius, or the creative mind) as a powerful natural force, overcoming all obstacles. Jacotot was advocating for his education system, which used exploration and discovery as a tool for learning, as opposed to the rote memorization which was the main method used in schools at the time.

Another author wrote,

> Already mechanical science has succeeded in binding down the wings of genius; and carpet manufacturers and fancy workers no longer consult the taste of artists, but apply to the kaleidoscope to supply them with new patterns. Are musical composers in future to be taught to take their inspiration from such an instrument as is now exhibiting?[1]

After a very successful opening in Paris, the attraction of the Componium soon waned. Honore de Balzac, the novelist,

---

[1] The Times, 20 May 1830, from *Componium* by Van Tiggelen p. 89

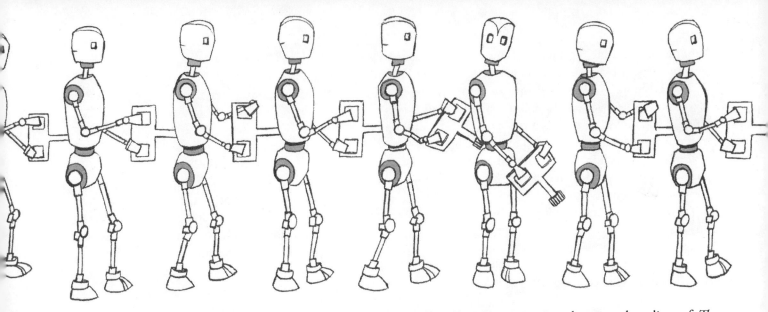

wrote with sarcasm about the gradually fading attraction "… these are perpetual Componiums, and one goes there regularly to seek for harmony indigestions."

The Eureka

The Componium worked by having a preselected set of pleasing elements, which were rearranged by a random process. This idea of random recombination of previously observed ideas was essentially how creativity was understood at the time, and the structure of many other devices was fundamentally based on exactly that same principle. It was applied to visual elements by David Brewster when he invented the kaleidoscope, and in one fascinating instance was even applied to writing poetry:

In 1677, John Peter published *Artificial Versifying, A New Way to Make Latin Verses* as a kind of entertainment for schoolboys. This explained a technique for composing Latin poetry automatically. Each verse was of an unvarying form:

Adjective Noun Adverb Verb Noun Adjective

The meter of each word was also fixed, forcing the line into metrically correct hexameter:

dactyl trochee iamb molossus dactyl trochee

For example, one of the lines the machine produced could be translated as "A gloomy castle sometimes shows a bright light."

John Clark, an inventor and printer from Bridgewater, England, began in 1830 to build a machine to carry out the steps of John Peter's process. He built a cabinet the size of a small bookcase that composed the poem while simultaneously playing "God Save the Queen." His device consisted of six turning cylinders, one for each of the six terms in the line of poetry. If it had simply displayed six words, it would have been regarded merely as a clever plaything. But Clark encoded the words using pins in such a way that they would cause individual letters to fall into place, apparently at random. This gave the impression that the machine was somehow composing the poem *letter by letter*, which was much more impressive. He deliberately fostered this illusion, writing in a letter to the editor of *The Athenaeum*, a monthly magazine:

> Permit me, as the constructor of the Eureka, or Machine for composing Hexameter Latin Verses, to make a few observations on its general principles, in reference to Dr. Nuttall's remarks, in your last week's paper. The machine is neither more nor less than a practical illustration of the law of evolution. The process of composition is not by words already formed, but from separate letters. This fact is perfectly obvious, although some spectators may probably have mistaken the effect for the cause—the result for the principle—which is that of kaleidoscopic evolution; and as an illustration of this principle it is that the machine is interesting—a principle affording a far greater scope of extension than has hitherto been attempted. The machine contains letters in alphabetical arrangement. Out of these, through the medium of numbers, rendered tangible by being expressed by indentures on wheel work, the instrument selects such as are requisite to form the verse conceived; the components of words suited to form hexameters being alone previously calculated, the harmonious combination of which will be found to be practically interminable.—Yours, &c. J. Clark. July 2, 1845.[2]

At the time of its exhibition (for one shilling at the Egyptian Hall in Piccadilly, London) the device attracted a lot of attention. Writers in the press at the time described the machine as "composing," "selecting," and "thinking." Writing Latin poetry was a common assignment in schools of the time, and William Thackeray joked in *Punch* magazine that "several double-barrelled Eurekas were ordered for Eton, Harrow, and Rugby."[3] One author wrote, "I do not see its immediate utility; but as something curious, it is, perhaps, entitled to take its place with

---

2   John Clark, *Atheneum*, 1845
3   Punch 9, 1845, p. 20

Babbage's Calculating machine, and inventions of that class."[4] In fact, Charles Babbage was familiar with the machine and with its inventor. William Hodgson, an economist, wrote, "It is truly a curious machine. Though I cannot say much for the sense of the verses.... The inventor spent fifteen years upon it, five years more than are needed to make a boy into a verse-making machine, and still less perfect. Clarke is a strange, simple-looking old man. Babbage said the other day that he was as great a curiosity as his machine."[5]

On its front face was inscribed the lines:

> Eternal truths of character sublime,
> Conceived in darkness, here shall be unroll'd;
> The mystery of number and of time
> Is here displayed in characters of gold.
> Transcribe each line composed by this machine,
> Record the fleeting thoughts as they arise;
> A line once lost will ne'er again be seen;
> A thought once flown perhaps forever flies.

Another point often mentioned in the tabloids was the combinatorial explosion of possible sentences. Over the course of a week, one journalist noted, the machine, if left running, would produce over 10,000 unique verses. "It becomes a torrent, poetry enough to make a person choke. It is like a snowy waste, where the unique, delicate snowflakes pile up to form mile after mile of mind-numbing sameness."

The device also included a kaleidoscope which produced a unique illustration to accompany each verse. The inventor was aware that both of these machines were performing analogous functions, that what he was building was just one of a class of "creative" machines. The Eureka has been maintained and can still be seen in the Records Office of Clarke's factory in Somerset.

The idea of a poetry-generating machine was a kind of running joke from this time period until the early twentieth century. For example, "The Poetry Machine" was a short story by Charles Barnard published in 1872. In this story, a young boy happens upon a poetry machine:

> He went up to the table and stood before the wonderful array of cranks, wheels, and levers. The machine was about three feet long and two feet wide and high. There was a clockwork attachment, with a weight that hung on a pulley under the table. It resembled a telegraph machine. There was a long ribbon of paper rolled on two wheels, and it had a marker, just as Morse's instrument has, to print the words. On one side were a number of stops or handles, with ivory heads, having curious words marked upon them. One was marked, "Serious," another, "Comic," another, "Serenades," and so on; one was marked, "Stopped rhymes," another, "Open rhymes," and there was one marked "Metre."

The boy generates poems without meter and with nothing but the rhymes as he learns how to operate the machine. The story serves as a parody of the kind of thoughtless poetry that was churned out for commercial jingles or greeting cards. Besides similar stories, "poetic machine" was used as a humorous metaphor for the poetry-making capacity in the poet's mind (especially for poets whose primary concern was making sure each pair of lines rhymed). All of these references assume that the reader will agree that simply "turning a crank" to generate poetry is an absurdity.

The Logical Piano

An *ontology* is an attempt to explain the world by systematically categorizing it. In the ontologies built during this time (following the suggestion of Aristotle) humans were characterized as "the logical animal." Logic was the highest form of mental activity; it was the main thing that divided us from the animals. The attempts to mechanize logical reasoning were thus seen as attempts to mechanize thought itself.

In the late 1700s, the Earl of Stanhope developed a device he called "The Demonstrator" for returning the results of a logical deduction. It was essentially a cleverly designed look-up table. This inspired Stanley Jevons to build the much more ambitious "Logical Piano" in 1869. Of its invention he wrote:

> As I awoke in the morning the sun was shining brightly into my room, there was a consciousness on my mind that I was the discoverer of the true logic of the future. I felt a delight such as one can seldom hope to feel. I remembered only too soon though how unworthy and weak an instrument I was for accomplishing so great a work and how hardly I could expect to do it.[6]

It was called a "piano" because it physically resembled a short upright piano. It had a tall case containing the mechanism, and below that it had an input keyboard, including a *finis* key that doubled as "Enter" and "reset."

Suppose one wanted to represent the fact that proposition A implies proposition B, and that proposition B implies proposition C. Given the input "A = AB and B = BC finis" typed into the keyboard, it would, through a system of levers and pulleys, return that the possible states include ABCD and ABC~D, but not, for example, A~BCD. (The symbol ~ means "not.")

However, the inventor of the Venn diagram, John Venn, pointed out that the same thing could be worked out on pencil

---

4 *Littell's Living Age*, Volume VII, p. 214
5 Life and Letters of William Ballantyne Hodgson, 1883, p. 52
6 C Black and R Konekamp, *Papers and Correspondence of William Stanley Jevons*, Volume 1, 'Journal of William Stanley Jevons' for 28 March 1866, MacMillan Press, 1973, p. 204

and paper without significantly more trouble. Jevons himself admitted that the piano wasn't of much use besides as a teaching device.[7] Venn's description of what is missing is a prescient description of what today we would call a programming language and compiler:

> I have no high estimate myself of the interest or importance of what are sometimes called logical machines, and this on two grounds. In the first place, it is very seldom that intricate logical calculations are practically forced upon us; it is rather we who look about for complicated examples in order to illustrate our rules and methods. In this respect logical calculations stand in marked contrast with those of mathematics....
>
> In the second place, it does not seem to me that any contrivances at present known or likely to be discovered really deserve the name of logical machines. It is but a very small part of the entire process, which goes to form a piece of reasoning, which they are capable of performing. For, if we begin from the beginning, that process would involve four tolerably distinct steps.
>
> There is, first, the statement of our data in accurate logical language. This step deserves to be reckoned, since the variations of popular language are so multitudinous, and often so vague and ambiguous, that they may need careful consideration before they can be reduced to form.
>
> Then, secondly, we have to throw these statements into a form fit for the engine to work with—in this case the reduction of each proposition to its elementary denials. It would task the energies of a machine to deal at once, say, with any of the premises employed even in the few examples here offered.
>
> Thirdly, there is the combination or further treatment of our premises after such reduction.
>
> Finally, the results have to be interpreted or read off. This last generally gives rise to much opening for skill and sagacity... I cannot see that any machine can hope to help us except in the third of these steps; so that it seems very doubtful whether anything of this sort really deserves the name of a logical engine.[8]

The logical piano, like the other devices described here, automated tasks that had previously been possible only in the mind. The artistic devices all shared an element of randomness, but here was a physical device that could reason and derive new truths from axioms in a completely deterministic way.

The Past's Unfinished Future

It is easy to think of these devices as dead ends. Only the simplest of them were ever developed beyond their prototype form, and today it would be much easier to program a general purpose computer than to try to bring any of these technologies forward. What they did, though, was to serve as an inspiration to begin thinking of machinery as something that could take on some of the roles that had previously been handled solely by the mind. What would have happened, though, if they had continued to be developed by a society that took them more seriously? A few possibilities come to mind:

- A self-driving camera on wheels that could capture and arrange photographic images into a multiple exposure, putting things and people together in some unexpected arrangement
- A tool that captures some of the elements of plot and story, allowing it to create its own fairytales
- A combination of a system for inventive sketching with a drawing automaton (like the one recently featured in the film *Hugo*) so that it could create original sketches each time it was run
- A device that creates an incredibly beautiful, haunting melody—the most beautiful melody anyone present has ever heard—but due to the design of the device each melody is played only once, and then lost forever.
- A divination system based on the principles of adjusting weights today found in neural networks, that could actually learn from experience to predict the future

There is a lot of untrodden ground here for speculation and research. It's easy to forget, as we imagine advanced Victorian robots, how far along those lines they actually had progressed.

Douglas Summers-Stay is a computer vision and robotics researcher for the Army Research Laboratory and a PhD student at the University of Maryland. He has recently published Machinamenta, *a book on the subject of creative machines before computers (available through Amazon.com for $8.45).*

Clark encoded the words using pins in such a way that they would cause individual letters to fall into place, apparently at random. This gave the impression that the machine was somehow composing the poem letter by letter.

---

7   Jevons was also the inventor of the economic theory of utility. See chapter IX on economics for more about Jevons.
8   *John Venn*, Symbolic Logic, 1881

FRENCHY &
THE FIRST CABARET
Illustration by Sarah Dungan

Frenchy & The Punk are a cabaret duo based out of New York that tours the steampunk circuit in a minivan and wins over crowds in music halls and bars with equal vigor. Sometimes their music is European folk-influenced, sometimes it's more punk. Sometimes, and most spectacularly, they put down their other instruments, take up drumsticks, and both attack their drum sets. Samantha Stephenson is the "Frenchy" of the pair (born in France despite her British name), and Scott Helland is the "Punk." After tabling next to them at TempleCon in Rhode Island, I made them answer some questions for us and our readers about life on the road.

STEAMPUNK MAGAZINE: Your project is certainly on the "punker" side of steampunk, while still clearly influenced by a wide range of folk and traditional sounds. What are your musical backgrounds that bring you to that place?

SCOTT HELLAND: Yeah, we definitely have an eclectic musical mix. While French chanson, Spanish guitar, and middle-eastern melodies are a big part of that mix, so is my experience in punk music. I started playing bass in bands during the hardcore explosion of the early 80s—I was already booking my band's shows, making flyers, promoting other bands and touring in the northeast before I had even

THE PUNK
STEAMPUNK FAERIE BAND

turned 16. One of the bands I was in, Outpatients, shared the stage with a ton of great bands from that era like Black Flag, Husker Du, Cro-Mags, GBH, and MDC. We went on to tour more extensively in the US, Canada, and Argentina. It was an amazing time and being a part of that early punk scene really shaped my musical path. I was also exposed to a lot of cool music growing up… my parents listened to the jazz and Gypsy jazz greats like Django Reinhardt and Stéphane Grappelli.

SAMANTHA STEPHENSON: What led me to where I am now musically started with dance. When living in England, I started dance at the age of four and continued into my 20s. I started playing piano at the age of eight and continued into my late teens. I was exposed to the cabaret, gypsy, and jazz sounds very early on so it became a natural part of my writing style. As a dancer, I gravitate to the more visceral type of music that inspires me to move and feel.

I grew up surrounded by music. My father played jazz and folk guitar and my mother sang—mostly French chanson Edith Piaf style. They didn't perform together or professionally, but it was a big part of their lives. When I was about six years old, we moved to Sao Paulo, Brazil and I became obsessed with the sounds of the Samba and Carnival of Rio. I found drumming hypnotizing. The drumming instrumentals that Scott and I started back in 2005 definitely stem from that early experience.

The Frenchy and the Punk sound has evolved a lot since we started playing together. I first joined Scott onstage as a percussionist at his live shows; he had been writing acoustic guitar instrumentals and I was providing some percussive accents to the tunes with tambourine, djembe, shakers, that type of thing. In 2006 I started writing some lyrics to the new riffs that he was coming up with, which eventually turned into the all-French CD, *Eternal Summer*, that we released in 2008 and the *At The Carnival Eclectique* CD which had the drumming and guitar instrumentals on it. As I got more and more involved in the song writing process, we started having more gypsy cabaret elements creeping into the tunes. We are both very fiery people so our music does have that punk energy to it, while still having a vintage vibe.

SPM: *Tell me about steampunk. You don't only play steampunk events, but you seem a good fit for the genre. How'd you get into it? What keeps you around? What other genres and scenes do you travel in?*

SCOTT: The steampunk community has really embraced us and it's been amazing. We've played many of the major steampunk festivals here in the US: World Steam Expo, Steampunk World's Fair, Steam-Con, AnachroCon, DragonCon's Time Travelers Ball, Steampunk City, and Dorian's Parlor. We started hearing about steampunk in 2005 or 2006 when we played at the Jeff Mach Wicked Faire in New Jersey. We do seem to fit in quite well, even with our eclectic sound, but that's the beauty of steampunk. The bands are very stylistically different from one another but there does seem to be that common thread of having a bit of a vintage vibe. We've played with folk, punk, gypsy, goth, metal, jazz, celtic, and renaissance bands. And besides the art and music festival circuit, we are big into the Faerie Festivals. We perform at them here in the US and in England. There is a lot of audience crossover between the two scenes. You could say we are the first cabaret steampunk faerie band.

SAMANTHA: Steampunk was filtering into the events that we were already doing. There is a lot of crossover from the Renaissance Faire, Faerie Festival crowd, and of course from the goth scene. We weren't playing in the goth scene, but we did appeal to a segment of that community because we write some material that touches on that energy.

Steampunk will have as many definitions as the number of people you ask. For me, it is a returning to the creative, the hands-on maker's spirit, bringing back the feeling that you can be a part of the process, warming up the coldness of the digital age with the aesthetic beauty of the past. Frenchy and the Punk is about connecting with your spirit so it is a similar objective but in a sonic way. We do a lot of faerie events too and probably for that very reason because of the playfulness. Faerie Festivals are more magical in nature but it is still within the fantastical.

SPM: *The two of you are pretty nomadic, from what I understand. What does it take to live your life that way? How often are you home?*

SAMANTHA: I traveled a lot as a child and always loved being on trains and ferries. I love the feeling of being in transit, the sense of adventure and discovering new places. We travel mostly in our van here in the US and we spend a lot of time on the road. We love to stop at random places along the way like the Czech Village in Iowa which ended up being the inspiration for the song "Vitame Vas" (Czech for "welcome") on *Happy Madness*. When we first starting touring we used to do a lot of sleeping in the van in parking lots and rest areas. I would use the outside water spigot, which I think are there for dogs, to wash my hair! Thankfully we did less and less of that as the years went by; we've met so many awesome people from touring that we have lots of places to stay now. We do still go to camping grounds though, on days off during the warm months, and the occasional hotel. When we toured in Europe last year we were able to stay with friends and family of mine, which is great, and we also stayed with people who set up the shows.

Life on the road takes a certain personality. It's definitely not for everyone, you really have to love it and not be one to need the comforts of home, the security of a paycheck, health insurance, or be adverse to risk taking! You have to be self-motivated because you're your own boss and at the end of the day you are responsible for everything. I think the fact that we are both so passionate about what we do, that passion really takes over… we don't have much desire for things that advertisers want people to think they have to have! We are pretty fixated on creating and playing our music and leading a creative life so we do whatever it takes to be able to do that. We do have a home base in NY. We can't have pets or plants since we're rarely there. Not being able to have a cat or a vegetable garden are the only things I miss.

SCOTT: To completely live this creative life we have to live very simply. Lots of little things make life on the road able to happen. Simple things like having a travel refrigerator in the van, eating healthy and cooking on the road, having very little attachment to things at home base and literally stopping to smell the flowers by the side of the road. Being true to yourself and being happy in the moment and loving what you do. We don't smoke and rarely drink, except for the cases of red wine Thursday through Wednesday… We don't have kids, or regular full time jobs. I guess our kids are the CDs and artwork we create. We make and sell our artwork on the road too: Samantha makes awesome fabric finger puppet bats and I do intricate pen-and-ink drawings. A lot of the artwork we do is on our CDs, shirts, stickers, and patches. Most of any money that is made basically goes right back into the band. Being in Frenchy and the Punk is a triple time job, but that is our choice and we like it that way. We love being on the road. We are away from our home base in NY at least 250 days of the year and we average over a 100 shows a year. We are a completely DIY band too. It's a way of life, and it's what I've been doing for the past 25 years. We feel blessed to be doing what we do.

SPM: *What is life on the convention and festival circuit like? Is there a culture of your fellow travelers, or is it fairly isolated?*

SCOTT: I feel like there is a little bit of both because we live in the extremes. We can spend a weekend at a convention with tons of people and then spend the next three days on the road driving to another gig and not really be in contact with anyone except for over the internet. But there are definitely bands that we see and share the stage with around the country that we get on very well with and are always happy to see. It's kind of an unwritten kinship. We see a lot of attendees and vendors at different events too, so it is like a family reunion. Some of the nicest people we meet are on the road and in this community. It's a tribe I guess.

SAMANTHA: People who play and vend at festivals are definitely a different breed, so yes, there is a culture. On the contrary, I think I would feel more isolated if I stayed in one place because we exist in a world full of the communities that we connect with. If we were in one place all the time that wouldn't be possible. But I agree with Scott, we can be isolated on the road and in what we are doing because we bear a heavy load and there's no one really to complain too. As difficult as it is, we wouldn't choose a different life so we just have to keep going forward no matter what. There is definitely a comfort in seeing familiar faces on the road and it's great to exchange travel tales. Even attendees travel quite a bit too, so we'll see fans in totally different places. That's especially true of the steampunk conventions. We saw people in Seattle when we played SteamCon that we knew from the east coast. It's the same when we do World Steam Expo in Michigan or DragonCon in Atlanta; we'll see people from all over. Then there are bands that we play with like The Men That Will Not Be Blamed For Nothing that we do shows with here and in the UK.

Time takes on a whole other feeling too; there is no routine so life is a lot less linear. We can get up and go whenever we want, the only set thing we have is when we have to be at a show. Our biggest constraint is financial—it's not easy to live a creative life, but I decided years ago that I would rather it be a financial struggle than live a lie and struggle spiritually. It's a choice I made. Now I just hope that our dedication, passion, persistence and perseverance will allow us to continue to do what we are doing for a very long time.

Frenchy & the Punk can be found online at frenchyandthepunk.com. Their latest albums are Hey Hey Cabaret, *released May 2012, and their drum and dance CD,* Elephant Uproar, *which ought to be out by the time you read this.*

Times Picayune
BOARD OF TRADE
RUE SOLIDARITY
RUE KANSAS
CLOSED
CLOSED
CLOSED
A.F.L.

The New Orleans General Strike of 1892 Laughs at Your Shitty Attempts to Divide the Working Class With Racism

*by Miriam Roček
(aka Steampunk Emma Goldman)
Illustration by William Petty*

You know what's key to a successful strike? Numbers. Well, a lot of things are key to a successful strike, but numbers are definitely on the list. So it's no surprise that the 1892 New Orleans General Strike looks formidable right off the bat: between the actual strikers and their families, half of the freaking 1892 population of New Orleans were participants. This wasn't some group of dudes being like "working conditions suck, let's strike! Hey, why is everyone either ignoring us, or, since this is the 19th century, beating twelve kinds of shit out of us?" This was literally half of the city of New Orleans standing up and calling bullshit on the city's Board of Trade and their tendency to stick their fingers in their ears and go "lalalala, I can't hear you" when unions tried to negotiate with them, which is a thing that can get really fucking annoying, as you will know if you have ever dealt with a petulant six year old, or a 19th century Board of Trade.

Before I really get into this, I want to make it clear just why this particular historical incident was so full of win. It was an anomaly. The striking (haha!) thing about the New Orleans General Strike of 1892 (spoiler!) was the refusal of the various unions and labor organizations to engage in racism or to divide along color lines, and the reason that is something worth talking about is because it was *unusual*. Unions of the 19th century were appallingly racist, using organized labor to keep people of color out of jobs they felt should be held by white people, even jobs that had, prior to the rise of organized labor, been traditionally held by black people. I want to make it clear that I do not intend this article to imply that the events we're going to be talking about in New Orleans in 1892 were an indication of some kind of lack of racism in the 19th century labor movement. I do mean to imply that the 1892 General Strike was pretty much the best thing ever, because it stood out like an awesome thing in a sea of racist bullshit. (Yes, exactly like that.) Anyway, to continue.

The whole thing started with three local unions, which happened to be fully racially integrated, one of them being predominantly black, deciding that they wanted better pay, and a shorter work day. Not the legendary 8 hour day, you understand, no, these guys were going for a 10 hour day. Now, it happens that all of these unions were affiliated with the AFL, the American Federation of Labor. New Orleans had, just that year (which was 1892, weren't you paying attention to the title?) seen a huge increase in union membership, and the AFL was actually pretty huge right then, which meant that having the AFL on your side was a definite Good Thing. Within New Orleans, the workers had formed something called the Amalgamated Council, an organization of unions that represented over 20,000 workers. Keep them in mind, we'll get back to them. But let me back up a bit here, because I sort of forgot to tell you guys what the political climate of the 19th century Deep South was in regards to the labor movement.

It. Was. Shit. Strikes tended to result in brutal retaliation and bloodshed, and Louisiana in particular had just a few years prior seen a horrific massacre of striking black sugar cane workers at Thibodaux, in what was aptly known as the Thibodaux Massacre, probably the single bloodiest conflict in American labor history, and woefully ignored by a lot of people (googling "Thibodaux Massacre" will get you less than one tenth the hits of "Haymarket Massacre" despite the second incident going by a lot of other names and having a lot fewer corpses) because, say it with me kids, **all the men, women, and children who were murdered at Thibodaux in cold blood foolishly failed to be white.** *That*'s no way to get into the history books, guys! So it was with that and countless other instances of racial violence and anti-union action all in mind that the New Orleans unions chose to act.

Oh, and before I move on to a far less rage/depression-inducing strike (the one this article is actually about), I want to share with you an irrelevant factoid about another failed strike by black agricultural workers in the late 19th century South. I couldn't rationally fit this factoid into any part of the article, so I'm just going to share it with you here and hope you appreciate it as much as I do. It seems, you see, that in 1880, black agricultural workers were on strike for a minimum wage of a dollar a day, and threatening to leave the state if they didn't get what they wanted. I bring this up, not because the strike was successful (it was broken, and the strikers jailed) but because it had the best slogan I have ever heard, "A Dollar a Day, or Kansas." See, it makes sense in context, since they were threatening to leave the state for better jobs elsewhere, but seriously, how awesome is "or Kansas" as a threat? I'm going to use it from now on. Hey dude next to me on the subway! Your balls do not need that much space. Stop man-sitting… or KANSAS! Hey internet troll. Don't troll me… OR KANSAS!

Try it sometime, I like it.

So. New Orleans, 1892. Like I said, there were a lot of issues being struck for, but the main one, the one that got everyone on board, was just the recognition of the legitimacy of unions, period. Yeah, the Triple Alliance (the name given to the three racially-integrated unions who started the whole thing, namely the Teamsters, the Scalesmen, and the Packers) wanted certain specific things; a preferential union shop, overtime pay, that ten hour day I was talking about, etc, but what brought the rest of the unions in was the fact that the city Board of Trade didn't just refuse to give them those things, it refused to fucking sit down and talk to them like grownups. During the first week, 3,000 workers (or about six percent of the people in New Orleans) were on strike, and **no negotiations took place.** The Board of Trade, who I'm going to start calling The Board because it sounds kinda like The Borg and therefore appropriately evil, was all "no, we will not talk to you about these issues, unions. Instead we will form a committee to raise money for 'defense,' ask the governor to send in the fucking state militia, and wait for the press to launch a series of really gross racist attacks."

So, with The Board being assholes, all the other unions in the city kind of sat back and went "huh. I wonder how well we'll do at being taken seriously if the city is basically announcing its intention to screw over unions, always, for no reason. Probably not well." Essentially, they realized that the Triple Alliance's problems were their problems. The Board made no secret of the fact that they basically thought unions in general were the enemy here, and, surprisingly enough, by doing that, they got a whole lot of other unions to be *their* enemies. I believe the term for that is "solidarity." And we're going to get back to that word, I just want to talk a little more about the strike first.

The Board had a cunning plan! Remember how there were three unions in the Triple Alliance? I mean, that makes sense; it's in the name. Well, only one of them was predominantly black; the other two were mostly white. So, what the hell, The Board said. We'll totally sit down and negotiate. With the white guys.

There have been countless times when exactly this happened, when the white workers were quick to say "sure! We never liked black people anyway." The Board made the offer because it was a strategy known to work; it was a great way of getting workers to turn on each other.

This time, though, the white unions responded with a resounding "are you fucking kidding me?" The white workers refused to talk to the Board in the absence of the black union, presumably asking The Board precisely which part of the concept of a Triple Alliance they didn't fucking understand. The two white unions returned to the picket lines, and solidarity was maintained. No one was going to be agreeing to any terms until those terms applied to everyone.

Despite the clear awesomeness of the strikers, there was not a lot of public support for this strike. Well… that's kind a misleading thing to say. With half the population of the city affiliated with either the Triple Alliance or the Amalgamated Council (remember them? Well, they weren't striking just yet, but were eagerly standing by with pom poms yelling "we love strikers! Go strikers!"), how much public support do they really need? They *were* the public, right? But the newspapers were pretty anti-strike. The *New Orleans Times-Picayune* in particular seems to have been hilariously pro-strikers' goals, but anti-any-means-by-which-such-goals-might-be-achieved. They were all "what? Why won't The Board just recognize that unions are a thing? Why is that a problem? Damnit, the Board should recognize unions!" while at the same time being like "going on strike? What are you people, *animals?* What is *wrong* with you?" It's also kind of remarkable how many of their headlines proclaim that the strike is totally over and done with, thanks for your time, we're so glad everyone's going back to work yay! that were incongruously published right when the strike was really getting going. I'm not sure if that was like, bad information, wishful thinking, or just a flat-out attempt to lie to the public. Possibly some combination of the three.

However, just a few short days after those incorrect headlines, the *New Orleans Times-Picayune*, which I am going to be shortening to *NOT-P*, even though that looks weird, was forced to publish the following headline:

THE STRIKE-
A Majority of Union Bodies Hold Meetings-
And Agree To Strike Whenever The Amalgamated Council Say the Word

That was November 1st, 1892. So now all we need was for the Amalgamated Council to say the word. Meanwhile, the *NOT-P* continued publishing almost daily reports that the strike was definitely over for *sure* this time, despite the fact that it totally wasn't, as well as occasional bits of strike-based "humor" in a little joke column called "Our Picayunes" which I guess would be the equivalent of a modern-day column called "Our Two Cents," and it's exactly as hokey and awful as that name suggests. For real, these jokes are painfully unfunny. Sample (for which I apologize, but it feels necessary for establishing context for the strike jokes): "seamless dresses are coming into use. If they seem less than the ball dresses that have been worn, they should not be allowed."

Big laffs. Anyway, the, and I use the term with hand-injuringly large finger-quotes (or just regular quotes, since I'm actually typing this, not speaking it) "humor" they had to offer on the strike was:

"There are many things in nature which are strikingly beautiful; but nature has not yet acquired the habit of going on strike." I guess the point of that one is that… strikes are bad? But not as bad as puns? Fuck you, *NOT-P*, if that pun is the best you can do.

"Man earns his bread by the sweat of his brow, until his brow strikes." I don't think I get this one. Maybe this one could actually be construed as pro-strike, if you assume that the "man" stands in for the bosses, and "his brow" for the workers? Am I reading too much into this? I'd love to know if people at the time found this shit funny. Was this like, the Stewart/Colbert of the day, or was it more like old Family Circus reprints? Only time-travel can tell!

"It is hoped that the hearse drivers will not strike. It would be awkward to make a man walk to his own funeral, and the coach drivers strike would be crowding the mourners." Okay, that one is actually worth a morbid chuckle. That is legitimately a little bit funny. Good job, *NOT-P*! Now, let's get back to your misleading coverage of the labor conflict.

By November 4th, *NOT-P* couldn't keep pretending the strike wasn't going to happen; the meetings with that committee The Board had formed were going nowhere, and the Amalgamated Council was on the verge of calling for a general strike, just like they kept saying they would. So the newspapers abruptly switched tactics. Instead of denying the existence of the strike, they started hand-wringing about what an awful thing this strike would be. Not because it would be bad for commerce; they'd been doing that for weeks. They started

> The Board of Trade, who I'm going to start calling The Board because it sounds kinda like The Borg and therefore appropriately evil, was all "no, we will not talk to you about these issues, unions. Instead we will form a committee to raise money for 'defense,' ask the governor to send in the fucking state militia, and wait for the press to launch a series of really gross racist attacks."

raising the possibility that the strikers would turn violent. Where they got this idea, it's not exactly clear. *NOT-P* actually tried to attribute it to the labor leaders themselves, but it didn't quote anyone, instead just said that the leaders "know" that it will be hard to stop 25,000 strikers from turning violent. Ominous, *NOT-P*. Very ominous.

The papers seemed to take it as a given that a strike of that many people would be violent. Just… well, because. Because 25,000 people! Because strike! **Because, damnit!** Because of reasons. And because, though they mostly (but not entirely) left this in the subtext, because a lot of the people on strike were black.

After a couple of false starts, the general strike began on November 8th. One half of the city was now officially on strike; black and white workers, almost all of the unionized work force of New Orleans. Forty-three different unions. Think about the fuss in your city the last time there was like, a transit strike. **That was one union—this was 43 and included the coal shovelers and the shoe salesmen. Deal with that.**

That violence everyone talked about kind of failed to materialize, though. What did happen was the almost complete shut-down of New Orleans. Stores were empty. Street cleaning stopped, as did gas service, leaving buildings dark. The lamp-trimmers responsible for keeping the streetlights going stopped working too, and eventually the streets went dark as well. Even firefighting stopped. *Performances at the opera house were suspended.* Shit was getting real.

But see, November 10th front page headlines were a bit odd. They talk about the governor calling for a militia, and the forming of a "volunteer police force" which sounds to me like a mob with badges, but I could be wrong (they might not have had badges). What those headlines don't mention is any reason why the city would *need* such things. They make absolutely no mention of strikers doing any of the violent things that the papers had been so worried about. Basically, they wanted this force to help protect scabs. Booooooo!

Meanwhile, though, there were people with something interesting to say about all this business. A newspaper called the *Boston Traveler* (a fairly radical journal) spoke up noting that it was really amazing how all of these white people were on strike in support of the black people in the Triple Alliance. Of course, they made the same points I did earlier in this article; this was a huge anomaly, and really only an example of people doing, for once, what they should have been doing always, but still, people were noticing an odd amount of unity and, dare we say it, solidarity.

Hey, can we talk about that word for a second? I said I was going to. Let me just break down really quickly how I think it should and should not be used:

Correct:
Black Activist: We're trying to get stuff done here, and the authorities are mistreating us!

White Activist: Hey, the authorities are mistreating those people of color. Let's show we won't tolerate that, and stand in solidarity with them against the authorities!
Everyone: Yay! Solidarity!

Incorrect:
Black Activist: I've noticed some racism within this movement.
White Activist: Why are you being so divisive? Where's your sense of solidarity?

Incorrect:
Female Activist: I've been sexually assaulted by a male participant in this movement.
Male Activist: Shhhhh! Where's your sense of solidarity?

(Any usage of the word that involves singing "Solidarity Forever" is also officially **Correct.**)

In the case of the New Orleans General Strike of 1892, it was definitely the correct kind, not the other kind, which is cool. And yes, the white union members were motivated by their fear that one day their own unions would have the same trouble getting recognized; this wasn't some kind of act of pure altruism, but see, in a way, I think that's what makes it awesome. Everyone, black and white, was just like "hey, I see we have mutual interests. Want to fight for them? In solidarity? Fuck yeah." They recognized that fighting those in power means not collaborating with those same people to oppress others. Good for them; they win the "basic human decency" award.

So with all this good-type-of-solidarity running amok, what could the authorities do? Cave to the demands of the strikers? Hmm, that sounds hard. How about racism! Racism would probably help, right? (This was the 19th century South. Racism was their fix for everything. People used racism to repair clothing, plug holes in roofs, and to get rid of aphid infestations. If you went to the doctor he would tell you "take two racisms and call me in the morning." If tech support had existed, and you called them up, the first thing they would say, instead of "have you tried rebooting?" would be "have you tried racism?" I'm not sure what the hell I'm talking about anymore, but what I'm saying is, these dudes were racist.)

So the papers started making all kinds of clearly racist accusations and implications about the strikers. The *New Orleans Times-Democrat* published articles about how the black strikers wanted to "take over the city," and the *NOT-P* ran a story about white women and school-children being insulted by "the blacks." Interestingly, the article doesn't actually say "white women and school-children." It just says "ladies and school-children" but it's pretty clear that, as is frequently the case, white is meant to be read as the default setting for human beings. No evidence whatsoever is provided that this incident actually took place, and, come on, if you were a black striker in 1892 in Louisiana, would you be more concerned with, like, trying to get recognition for your union, or for some reason insulting passing white school-children? Who bothers to insult school-children anyway? It hardly seems worth it. The accusation that white "ladies" had been insulted was clearly a dog-whistle, intended to stir up animosity and violence against the black strikers.

Here's the thing about racist, mob-inciting dog-whistles in the 19th century South: they usually worked. This one… didn't. No mobs formed. No violence happened on the picket lines either, to the point where the authorities were actually kind of confused about it. The mayor sent out a call for those volunteer "deputies" to fight the strikers… and less than sixty people turned up. The mayor banned public gatherings, which is a totally awesome and constitutional thing to do, especially when there's no sign of violence at all, and The Board finally convinced the Governor to send in the militia for… some reason, but when they showed up and found exactly zero people behaving in ways worthy of militiaing, they just kind of turned around and went home.

It was awesome. All these racist white people just *waiting* for the black strikers to get violent so they could have some reason to shoot them, and a whole bunch of nothing happening, and The Board and everyone else just panicking about all the violence they were *sure* was going to happen any second now, because I mean, come *on*, there were *black people*, and *unions*, but just… nothing.

Finally, The Board agreed to sit down and enter binding arbitration with the unions. And… well, here's where the story gets kind of… not sad, but complicated. Because you really want this story to have a happy ending; I mean, I do, and I assume you do too, or you're probably an asshole, but it's actually a mixed bag. A lot of workers ended up with higher pay, and shorter hours, but they did not get the union shop they wanted, and none of the unions of the Triple Alliance ended up gaining recognition. At the time, a lot of people saw this as a victory for the unions; it certainly increased union membership substantially in Louisiana, but in the end, the unions lost a lot, too.

Here's what I say, though. This was less a victory of organized labor than it was a victory against racism. The fact that the unions didn't gain recognition sucks, but the fact that they were unwilling to divide along racial lines, even if it would have allowed some of them to get that recognition is awesome. This isn't a story about unions winning in a fight against the bosses; this is a story about a multi-racial group refusing to let their opponents use racism to incite violence or divide them in their struggles, and for that, I think the New Orleans General Strike of 1892 does represent a victory.

And on that note: Solidarity forever (or Kansas)!

Scarabs
Katie Casey

"This is the saddest thing I have ever seen," Amelia said, and George's face fell; coming from her, that was quite a declaration. He recovered quickly though, taking the gadget from her hand and exhibiting it proudly.

"Wait until you see what it does," he said, and began to describe how to turn it on.

Amelia's mind wandered, losing track of her brother's explanation to look at the designs he'd shown her when he'd proposed the idea, based on pictures ripped from the pages of Lady Eleanor's catalogues. Those scarabs—that's what the little devices were called, bug-like as they were—were smooth and round and decorated with gems, designed to be cleverly concealed as a hat decoration, the syringe hidden in the pin, or worn on a chain like a pocket watch. The thing in George's hand was… Amelia didn't know the word for that shape, even, with so many scraps of metal forced together at odd obtuse angles. Almost a circle, perhaps. Aspiring towards it. Neat lines of tiny rivets dotted its surface, with two larger ones near the top playing at eyes. While the ones in the catalogues folded neatly in on themselves, this one had sharp legs jutting from its body, and they shook slightly as it stood in George's hand, as if uncertain if it could hold its own weight.

"How does it work, then?" Amelia asked, though she wanted nothing more than to return to bed.

"He," George corrected. And then, when Amelia raised an eyebrow: "It's very important that you, well, make friends with him. All the articles said so—you have to build a sort of rapport, you see, or…" He trailed off, as Amelia's gaze wandered away again. "It could be a she, if you prefer," he added, a bit sheepishly.

The scarabs cost a small fortune, but were, George had assured her, easy enough to replicate; their cousin was a clockmaker, and had agreed to help assemble one from spare parts. There had been some dodgy moments; George had "borrowed" Lady Eleanor's scarab one evening to see how it was put together, and they had accidentally detached one of its legs. Amelia knew he had used all of what little free time he had in the evenings to put the gift together, and that he now owed their cousin a dozen favors and probably quite a few drinks, but grateful as she was for his concern, she couldn't bring herself to be enthused by the crawling little device. "How does *he* work, then?"

George flipped the scarab over—its little mismatched legs wobbled in protest—and pried open a panel, revealing an inside very much like a watch, but with a half-dozen minuscule keys next to a matching set of tiny clock faces. "You can set it to different times in the day; each key has a different task. For example, this one—" he pointed, though Amelia made no effort to sit up and look, "is for waking up." He twisted it, and set the thing back on the table; a moment later, it whirred into life.

"Wake up!" it sang, in so tinny a voice that Amelia wasn't sure she would have heard the words at all had her brother not just told her what they were. It was quiet and airy and altogether unconvincing; however, a moment later, the thing teetered on its legs across the table and jumped onto Amelia's lap, and proceeded to climb her dress.

"It won't stop until you get out of bed, see?" George said, looking entirely pleased with himself as he snatched it off her shoulder and turned the key again, stilling the thing. Amelia shivered, smoothing out her skirt where the scarab had rumpled it, as George explained the functions of the other clock faces—one to remind her to eat, one to remind her to bathe, one that simply made reassuring little whistles. "And this one is for the medicine," he said, indicating the center key. He looked up at Amelia, hopeful, and she shifted uncomfortably in her seat. "Shall we try it now?"

Amelia considered saying no—begging off, perhaps, on the grounds that she had quite a bit of work left to do that evening, and as much as she appreciated his gift, she thought that maybe it would be best to wait—but George looked so terribly concerned and excited all at once that she could not bring herself to say it. When Amelia offered no protest, he turned the center key. While the scarab had been somewhat spastic in wake-up mode, it moved slowly, almost cautiously, when George set it on her lap for the second time.

"You have to hold your hand out," he explained, demonstrating with his own hand resting palm-up on his knee. Amelia did the same, and the scarab teetered into her hand. The feeling of the legs skating across her palm made her shiver again, though she held her hand steady. A syringe emerged from one of the legs, and as Amelia gasped in alarm the scarab pushed the needle into her wrist. A moment later it was out, the scarab folding itself up again, and Amelia shuddered.

George put a gentle hand on her knee, then resumed his eager explanation. "I'll set it for every morning at breakfast time. You have to take the medicine every day, or it won't work properly. It needs to be refilled and wound up again every week or so, so I can do that when I come to call on Sundays. If you have any trouble before then, you could come by the house—but it should be fine, of course," he added quickly, seeing Amelia's alarm. He smiled at her, and took the scarab from her lap, checking the positions of the gears once more before looking at his watch. "I have to run, his Lordship is having guests for dinner."

Amelia pulled herself to her feet as George put on his coat and hat and leaned in to kiss her on the cheek. He set the scarab down on her table, where it sank into the small mountain of unfinished sewing.

It sank further when George left and Amelia fell back into her seat, pulling out two pieces of a shirt and her sewing kit and beginning to work. She put a few stitches, but had hardly made it two inches before the thread became tangled, and tugging at it in alarm only made the knot tighter. She started to cut the stitches out, then put her scissors down and picked at the thread, her attention wandering to the pile of unfinished work on her table, and the much smaller pile of finished ones draped across the back of a chair; she gave up at pulling out the stitches entirely as the thought of failing to produce enough shirts to pay the cost of her rent made her stomach turn. She shook her head to vanquish the thought and re-threaded her needle, lips pursed as she missed the eye once and then again before finally getting the thread through, and started again. She had only just finished the first shirt and was staring forlornly at the second, calculating the likelihood of finishing it in the remaining hours of good light, when a noise from the bottom of the pile of fabric nearly caused her to fall out of her seat.

The scarab forced itself across the table, dragging a sleeve behind it, and whirring: "Diiiiiiir," which Amelia understood to mean "Dinner" only after it had repeated the odd sound a dozen times, with increasing agitation, and walked itself right off the table.

"Oh, shut up!" she cried, and kicked it. It slid to a halt against the wall and waved its legs just a moment longer before closing up again. Amelia watched it for a moment, then sighed.

"Fine. If you insist. Dinner."

⁂

"Wake up!"

During the first week it had taken the poor scarab an hour of scuttling around Amelia's blankets to urge her out of bed, but she had quickly established a routine; she sat up, groped around until she found the thing, and threw it across the small room, where it landed against the wall with a very satisfying clang. She thought about going back to sleep—it would start insisting she eat in an hour anyway, and she saw little reason to be up before then—but her heart was still beating fast from the effort of chucking the scarab, and she found herself too awake to stay in bed.

The scarab righted itself and scuttled across the room as Amelia washed her face and dressed. It stopped when it reached the table, curling up by the leg which it seemed to have claimed as its preferred place for waiting. She eyed it warily—it was going to start chirping at her eventually, she knew, and she'd been ignoring its insistence that she eat for the past day and a half, too daunted by the prospect of going out to buy food. In previous such dire straits George had volunteered to do her shopping for her on his day off, knowing how difficult it was for her to be out and about when she was in one of her moods, but she blushed at the thought of asking again after he'd been so confident his little device would cure her of her melancholy. She wandered over to the window and peered outside. It was an unabashedly cheerful day, with the sun shining and a number of picturesque white clouds. Amelia took a step towards her bed, thinking of just crawling back into it, and stopped herself resolutely mid-stride, instead turning to pick up the scarab.

"I ought to just be able to send you to get things," she told it, running her finger over its rivets where it was curled in her pocket.

It took Amelia a while longer to feel prepared to step outside; she adjusted her hat on her head and re-tied the ribbon under her chin about a half-dozen times, stared at her empty cupboard deciding what she would need, patting her pocket to ensure that the scarab did not create an unsightly lump, and paced anxiously back and forth a few times for good measure. When the scarab buzzed again, announcing breakfast time, she twitched so violently that her carefully positioned hat fell to the side again, and she yanked the device out of her pocket and turned it off with shaky fingers before slamming it onto the table. She considered leaving it—serves it right, she thought, for startling her like that—and then scooped it up again, shoved it into her pocket, and set out before she could change her mind, closing the door loudly behind her and flinching again at the sound.

She intended to make it a typical day of shopping: go straight to the grocer's, buy just a few basics, and hurry home. But walking through the market on the way there, she passed a stall selling fresh vegetables, and hesitated in her beeline to the shop. The crates set up were piled high, and Amelia found herself wandering over, the ingredients list of her favorite stew drifting to mind as she turned a particularly lovely tomato over in her hand.

She was on her way home from the shop when a voice stopped her: "Amelia!"

It was a neighbor, a friendly woman named Sarah with whom Amelia sometimes visited. "I haven't seen you in ages! Your flat is always so quiet, I was beginning to wonder if you had moved and not told me. You look well!"

Amelia's mind immediately flew to the stain she knew was on her skirt, the mess she imagined her hair must be under her hat, and wondered if Sarah was making fun of her. "Thank you," she muttered. She closed her hand tightly around the scarab in her pocket, the hinge of one of its legs biting into her hand as she searched around for something to say. "Ah… Lovely weather, isn't it?"

"Beautiful," Sarah agreed pleasantly. "I was thinking of going for a stroll in the park later. You always run into all sorts of people out enjoying the sun, on a day like this."

The thought of running into anyone else made Amelia long for the quiet of her apartment. "Sounds lovely," she said, wondering if she ought to preempt any invitation to join Sarah with some kind of excuse. Nothing came immediately to mind, and she felt her cheeks growing warm. "Well I should, um…" she began quietly, but Sarah seemed not to hear her.

"You should come by for tea sometime," Sarah interrupted. "It would be lovely to catch up a bit."

"It would," Amelia responded noncommittally.

"Call by anytime, then," Sarah said cheerfully. "I really should be off, though. It was nice to see you again, Amelia! Do take care!" She waved goodbye and hurried on to her own errands, and Amelia exhaled with relief, though she kept a tight grip on the scarab as she walked home.

She left her groceries on the floor and sat on her bed for a moment, then pulled the scarab from her pocket. She turned the key which started its tinny little voice whirring, like a bumblebee mumbling. It was a useful distraction, turning her attention to the noise, which sounded for all the world like it might have been saying sympathetic things had it been capable of forming words. When it ran out, she sat a moment longer, and then set to work on dinner, her mood greatly improved by the promise of stew.

The day before George's weekly visit, Amelia realized that she faced a dilemma. The scarab was scuttling around her table as she ate her breakfast, and the many dings and dents in its surface caught the sunlight, revealing the abuse the poor thing suffered. George couldn't know that she periodically vented her frustrations on his gadget; the very idea was mortifying. But just as quickly, with a small flutter of excitement, she concocted a solution.

She had been cleaning her little flat lately, more thoroughly than she had in ages. It was in part an effort to avoid the depressing pile of sewing through which she was still picking, and in part because the scarab had proved adept at walking into every pile of rubbish or dust or untidy items in the room and then screaming encouragement from beneath, a habit even more annoying than its insistence that she get out of bed in the morning. She found herself cleaning up after it whenever she found it trapped under a discarded newspaper or tangled in some scrap fabric. She wondered if it did it on purpose; one small task of tidying seemed to lead to another, and the room felt a bit more cheerful with the bed made and things put away. Just a few days previously, it had somehow worked its way into the middle of an untidy stack of old paintings, forgotten under the bed along with a box of paints. Amelia had sneezed as she reached under the bed to retrieve the scarab and the paintings; dust obscured the colors, and water from some leak or spill had stuck the pages together. Now, she reached back under the bed for the paint box. George had given the paints to her for Christmas a few years ago, remembering her endless childhood scribbling. She had enjoyed them, painting a few little landscapes and portraits as gifts, but had almost forgotten them in the past several months.

Amelia pushed the pile of half-sewn shirts from the table and carefully removed each color, lining them up one next to the other. Some had dried out, and a shade of dull yellow had spilled across the bottom of the box and stuck several brushes together. She sorted them carefully across the table, pushing the ones too dried to use into a corner and arranging the rest in a row, then setting the brushes in front of them, as delicately as if they were feathers. She considered the arrangement for several silent moments, before removing the scarab from its spot under the table and examining the dents. There was no correcting them, of course, but painted in the right way she imagined they might be hidden, with George's attention too distracted by the colors to see the damage she'd done to his thoughtful gift.

She picked a brush and dipped it into a pot of green, and turned the scarab over once more before picking a starting point above one of its legs and coloring the metal in tiny strokes. She worked slowly at first, stopping often to hold the device up to the window and frown as she studied the way the light fell on its surface, but as the patch of green spread across the scarab she began painting more confidently. She imagined the dents as places where gems might be set in one of the scarabs from the catalogues, and filled them with rich colors, ringed in textured orange, the closest she could make to gold. When she switched to a dark blue, her brush slipped, the color slashing through and seeping into the paint around it, and she cursed herself for her sloppiness, her throat tightening as her thoughts spiraled away, loud and angry at her ungrateful damage to the scarab to begin with, that she even needed such a thing to simply be well. The scarab clicked, its legs moving against her palm, and she took a deep breath, and blew on the spot with the mistake, until it dried enough that she could paint over it.

She paused in her painting only when the scarab began kicking, demanding to administer her medication; it left blue footprints on her palm as it went.

"Diiiir!" the scarab buzzed as George refilled its store of medication and Amelia refilled his cup of tea. He raised an eyebrow.

"That's not right; it's hardly noon," he said, frowning at its mistake.

"It's been getting earlier every day," Amelia admitted, hoping her demure downward glance hid her theory that nearly three weeks of abuse—less recently, though she still found occasion to throw the scarab—had disrupted its many clocks. She sat as George fixed it, either oblivious or graciously unwilling to comment.

"It looks beautiful," he said again, putting it down on the table and admiring the field of flowers that had bloomed across the scarab's back as he sipped his tea.

"He," Amelia corrected playfully, though her brother had given up insisting that she name the thing after her last visit, when she'd coaxed him into admitting how ridiculous Lady Eleanor looked, conversing in hushed whispers with her hat pin. George smiled at her.

"It's good to see you happy again," he said softly.

The urge came up—familiar, quietly angry—to throw her teacup against the wall, to hear the shatter and splash, to insist that she was *not*. She took a breath, and let it pass. She had begun painting the walls when she'd finished the scarab—more cheerful flowers, but stranger things too—she had painted an intricate spider web in one corner, the strands spinning out into an abstract pattern of darkly colored swirls that pirouetted across the top of the wall. If her brother had found it unusual, he had not said, and she found herself grateful.

She nodded and took a drink of her tea, and he seemed satisfied enough. "You know the pharmacist on Flint Street?" he said, pressing a little switch on the side of the scarab so the vial which held her medication popped out. "I've told him he can start expecting you to come in for the refills yourself. On my account still, of course, until… well, the piecework is improving, isn't it?"

He was already standing up, his smile somewhat pained. She nodded again.

"Great," he said, as she stood up as well, and to her surprise he swept her into a hug. "I'm so proud of you, Amelia," he said, releasing her. "I'm not sure I'll be able to call by next week; I'll send word if I can."

"Alright," Amelia said. George had returned the scarab to the table; it skittered up to where she rested her hand, crawled up her wrist, and wrapped its legs around her arm in an odd copy of George's hug. "Have a good night, George,"

"You as well," he said, with one more affectionate look, and left.

Amelia stood for a while, unable to think anything in particular. She shuffled back to her chair, and stitched sleeves onto shirts until the light became too poor, using a rough edge of the scarab to trim the thread.

"Diir," it reminded her eventually, removing itself from her arm.

She put aside her work, stretching out her sore fingers, and considered the little device. "Perhaps tomorrow we'll ask if there are any jobs open, when we're in town," she told it. The thought made her stomach turn, but held itself in place all the same. "You would have to be very quiet, though."

It folded up its legs, which Amelia took to be a gesture of agreement, and she began to make plans as she cooked herself supper, for the first time in quite a while. ✣

INVENTING STEAMPUNK FASHION WITH PAPER DOLLS

Written and illustrated by Charlotte Whatley

Are you wondering what to wear to a convention? Invited to a steampunk wedding and haven't a clue much less a cog? Tired of tutus and top hats? And what *is* it with all the goggles? Designing your steampunk look does not need to be any more difficult than figuring out what a time machine does. Start with your imagination. For the sake of clarity, let's define *steampunk fashion*. Dissect it into understandable parts.

Steam comes from the idea that most inventions of the modern world could have been created using technology available in the expanding steam-powered Victorian Era (about 1840 to 1910).

Punk: the fuse for lighting cannons. **Punk:** the downtrodden and dirty street urchins. **Punk:** the aggressive, rebellious, do-it-yourself ethic. We are a community of mechanical magicians enchanted by the blend of history's fiction, fact, and fancy and are smitten with the mysteries of the imagination yet to be.

Fashion: The vogue. The now. The decorated coverings that tell the world who we want them to see.

Put these things together and you have a great start on creating interesting clothing designs. Although the possibilities may not actually be endless, blazing a trail along a steampunk Mobius strip of invention is tantalizing. This is my creative process for designing. May it serve you, as well.

1. **Have fun.** Whatley's Cardinal Rule of Life in General.
2. **Create or choose a character.**
3. **Know what the character does.** This will help determine what they need and like to wear.
4. **Research** the historic time period for ideas.
5. **Research** contemporary high fashion for ideas.
6. **Sketch** your designs. I use Clearprint 1000H Tracing Vellum, as pencil erases easily and it can be inked or transferred to heavier Bristol vellum. Additional costume designs are made by tracing the finished figure.
7. **Incorporate some Victorian elements** to your design.
8. **Incorporate some contemporary elements.**
9. **Add details** that are recognized as "steampunk" to the general public, such as gears, goggles, top hats, spats, etc. But use them *creatively*. Consider how they function as part of the design. This step is actually optional, in my opinion. Like poodle skirts from the 50s, not everybody wore them. Same applies to the aforementioned "steampunk" details.
10. **Remember: Tell it like it never was.**

PROJECT: Creating a Steampunk George Sand Paper Doll Costume

I have given you a list of my guidelines. Now follow my applied process... and see how I did it.

1. **HAVE FUN.** Whatley's Cardinal Rule to life in general.
2. **CREATE A CHARACTER:** In this instance, we are using George Sand, an actual famous female French writer of the mid 19th century, in order to explore more options of steampunk clothing for women than currently available.
3. **HISTORIC RESEARCH:** We discover she was paramour to Chopin, infamous for wearing men's clothing though not all the time. Though she claimed it was for comfort and economics, she could not deny the attention she received helped her break into circles previously exclusive only to men. She may or may not have had lesbian affairs, but certainly was fascinating to many men.
4. **CONTEMPORARY RESEARCH:** Paris Vogue, Fall 2010. Runway designer fashions inspired by 19th century dandies.
5. **DESIGNING:** The doll is *not* a portrait of George Sand, although her hairstyle has a similar look. The underwear is based on a historic corset illustration. I found no information indicating she wore men's underwear, too, but I doubt it. I'm going to say she had a secret vavoom sex life and wearing an Iron Maidenform for fun could be plausible (or maybe it was made out of satin and just stitched and appliquéd to look like gears and rivets.) The dandy costume is based on an illustration of her circa 1850. I was amazed at how contemporary (and cute) it looked. After drawing the basic slim pants and full skirted jacket, I added the details. Here is an excellent point for aspiring steampunk fashion designers: **details.** Be lavish with them. Victorians loved them. Fussy braids, laces, lots of buttons, creative stitching, beading and embroidery. Go wild. It is also here that I like to add the steampunk elements... with a generous sense of humor. I love visual puns. Notice her "stovepipe top hat" is really a stovepipe. Buttons on her jacket are replaced with cogs and chains that are functional as closures as well as decoration.
6. **YOUR ASSIGNMENT:** Using tracing paper, trace the outlines of costume in pencil. Now make just **one** simple change. Sleeves? Coat skirt? Pants width? Next, add something iconically steampunk, such as goggles. Lay the tracing over the doll to make sure it will fit and adjust if necessary. Ink it and photocopy it for reference. You may wish to color it. Repeat the entire process until you are pleased with the results and have several variations. You are now a steampunk designer. Have fun!

Charlotte Whatley is a paper doll artist with a background in Historic Costume, Museum Collections, and Fashion Design. Her award winning wearable art designs have appeared in several exhibitions and runway shows. She is the author of a unique paperdoll book, A Steampunk Tale Of The Curious Canine, His Best Friend And The Lady Who Flew, *published by Paper Studio, and Charlotte's paper dolls may be purchased through her website: charlottewhatleypaperdolls.com*

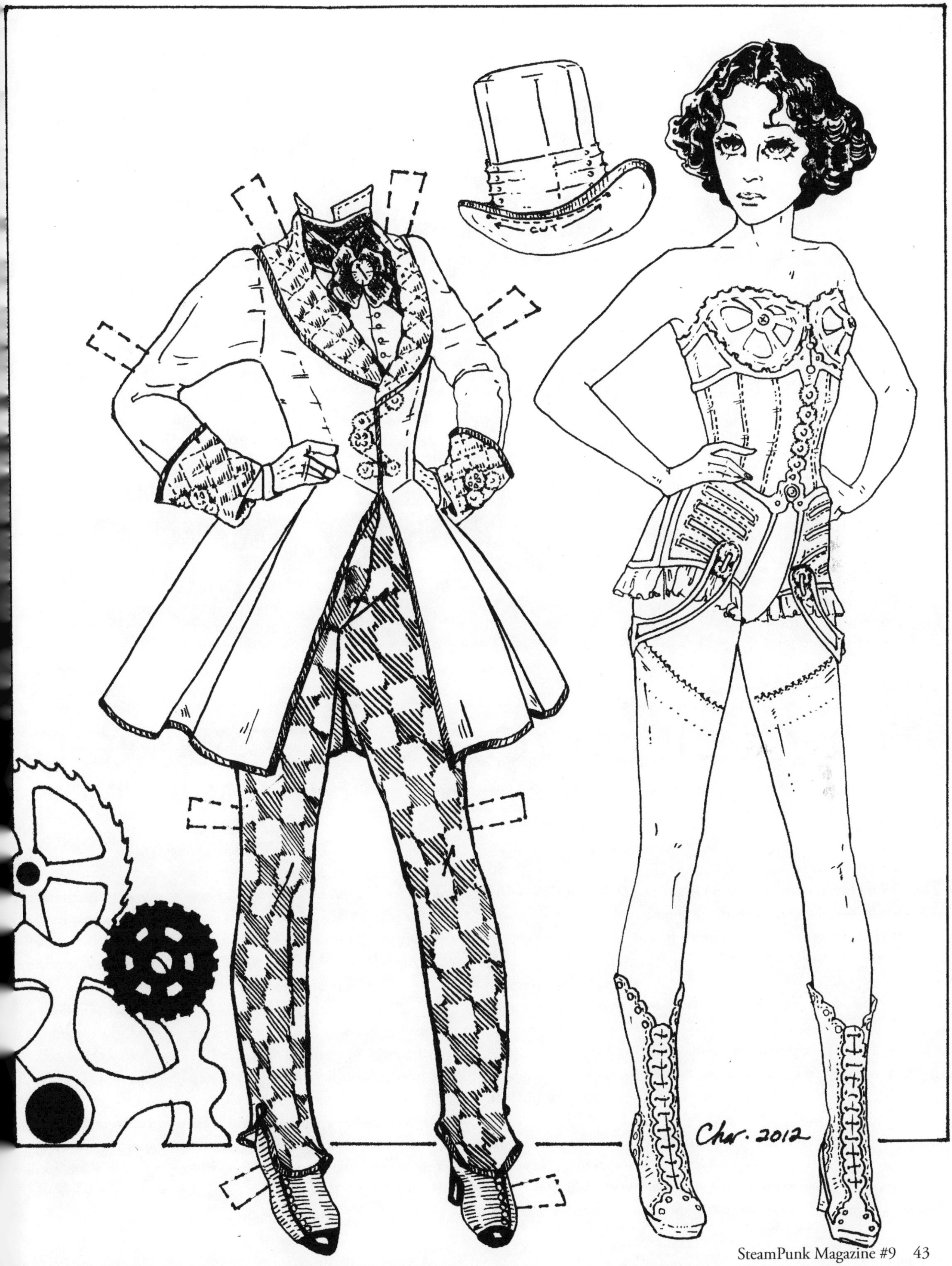
CUT
Char. 2012

NO CLOCKWORK REQUIRED
AN INTERVIEW WITH THE VAGABONDS

By Kendra Saunders
Illustration by Sidney Eileen

It's a cold Sunday afternoon in Salem, Massachusetts, which adds to the witchy aspect of the famous city as I meet up with several members of the steampunk group, The Vagabonds, for an adventure and interview. As we march over bricks and old streets, the pace varies between the more meandering steps of group members Eva and Adie, and the commanding, energetic steps of leader Locke Valor. Sometimes he gets far enough ahead that he circles back, and after even the earliest stages of our interview, it's easy to see that his mind and words move just as quickly as his feet.

The Vagabonds are a colorful group, in both a literal and figurative sense. Their members originate from states as far apart as Massachusetts to Michigan and Florida. Eva is soft spoken and offers advice and invaluable information about a plethora of subjects from sewing and costume customization to the importance of a sewing kit. Adie, the youngest of the group, is willowy and pretty and wears her pink hair in such a straightforward manner that you find yourself questioning your preconceived knowledge of natural hair colors.

After discovering that our restaurant of choice is closed for the day, we descend upon Salem's Beer Works and order our meal. We discuss their characters (Eva is the witchy cook who keeps accidently poisoning people, Adie is their gentle guinea pig who has a crush on Locke, and Locke is their mad scientist/paranormal detective leader with an eidetic memory) before diving into a conversation that covers everything from goggles to glue guns and back around.

STEAMPUNK MAGAZINE: *How long ago did you start making the vlogs?*

LOCKE VALOR: Wow, it's been about four years, maybe five.

SPM: *Why did you originally decide to use the video media to share your stories with fans?*

LOCKE: Because it wasn't really being used by a lot of people at the time. I wasn't the first, but one of the first to do actual vlogs.

SPM: *Was youtube around at the time?*

LOCKE: Oh yeah, youtube has been around a long time. Actually, I started releasing them on myspace first but myspace's graphics weren't as good. At the time of my first vlog, I created it as an intro for my character and placed it on myspace for the steampunk community. After a while I stopped posting them there and used youtube as the main platform. [As well as their website, www.vagabondsteam.net]

SPM: *How long does it usually take from beginning to end, not including storyboarding or story planning, to create one of your vlogs?*

LOCKE: When I first started, it didn't take too long, usually an hour or so, because I didn't use scripts. I just did it off the top of my head. The old ones are rougher. Now it takes maybe a half hour to film, but it's the editing that's a different story. Editing can go anywhere between ten minutes to five hours.

SPM: *And you do all of the editing yourself, on your computer?*

LOCKE: Yes. Any software works. There are free green screen programs you can use, there's free 3D programs. Windows Movie Maker can do exactly what you need it to do, including trimming clips, and it's a good beginner tool. You don't need anyone to walk you through it, it's basic enough to figure out with a little time. You can download that for free. Some people use iMovie, for Mac. Gimp is a good starter for pictures.

SPM: *How long have you been using green screen?*

LOCKE: Actually I did some green screen work back then, too. But the last couple vlogs I've used green screen on. It's getting better and better every time I use it. We're working slowly and secretly on a series that's going to be a completely different style that we hope hasn't been done yet. And we're thinking about making a podcast, like old radio dramatizations.

SPM: *What equipment do you use for the videos?*

LOCKE: Canon Powershot Sd1100 IS. Two basic desk lamps. Sometimes we'll use candles, it depends on the effect we want. For white balance, we just use pieces of paper. Our green screen is a table cloth. You can pick those up at Target or even a dollar store.

SPM: *What are some of the main changes you've made since beginning the vlogs?*

LOCKE: Scripting. We write out more of our story beforehand. We mostly did improv before. We use better lighting and better programming. Better camera too, because at one time it was webcam, and as we all recall, it had a mic issue. It was harder to sync it up. Our story is that we threw it off the ship.

SPM: *What are your suggestions for potential vloggers on a tight budget?*

LOCKE: Script. Work on a script... any vlogger out there that's known... they script whatever they say beforehand. Don't worry about jump cuts. Slow down between your talking, because if you mess up, it'll be hard to connect shots. Work on vlogging techniques. Any camera will work, even the worst webcam, but lighting is a factor. A dark room should be done in post, because if you do it during, it'll make the shot too dark and pixelated.

SPM: *The cool thing about what you just said is that these aren't expensive things. It reminds me of when Robert Rodriguez said he actually likes to work on a shoe-string budget because it forces him to be more creative.*

LOCKE: I'm actually a big fan of his style. He's a one man film crew, because he does everything.

SPM: *Switching now to the more general stuff, what would you say are the five most important items for a steampunk to bring to any convention?*

EVA: Costumes. Props, including weapons. Or, in my case, spoons and forks. [Her character is a cook]...

LOCKE: Cleaning supplies. Don't be a smelly one! Well, I guess that's optional. Money. Have your set money and have your backup money, just in case. You never know when unseen expenses could come up.

EVA: Cash, because you never know when you won't be able to use a credit card.

ADIE: Things to fix your costumes. Repair kits.

SPM: *What would be a basic repair kit that everyone should have with them?*

LOCKE: A sewing kit, glue gun with hot glue sticks, and glues that don't smell. Liquid paints, not spraypaint, because hotels don't allow spraypaint. You would have to take spraypaint outside. We personally use Folk Art paints.

SPM: *Now about costumes. Give us some tips on what's necessary for a steampunk in the clothing department.*

LOCKE: Goggles are not required. Goggles don't make a steampunk. Steampunk makes itself. To the public, when you think steampunk, you think goggles or gears on your clothes. A lot of people will go against that now and say you shouldn't even have them on your outfit. If you like goggles, wear them, but find a purpose for them. If you feel that your character would wear it, you can use it. I say, create your character first.

SPM: *That's a lot like what writers do. They dress their characters according to where the character came from or what they do. So people should pick things for their outfit that go with their character? Where should people look for inspiration?*

LOCKE: You don't have to have a character, that's optional, but most steampunks do. And with steampunk, you don't have to be in one world, you have multiple dimensions to interact with. Everything works. There's youtube stuff, there's books. You can look up other characters you like, Sherlock Holmes for instance. It depends on what you want your character to be.

Actually the traditional sheets of character creation, like from D&D, can help you. And you don't have to be a captain or an engineer. You can be anything. You could be a store front owner, or a street sweeper. Think how epic it would be to see a street sweeper... there's tons of stuff you can do. The more things you create, the better it will look. [Locke launches into a tangent about street sweepers and the possibilities with such a character, until all of us have to push him back on course.]

SPM: *And maybe add some of yourself to it?*

LOCKE: Yeah, you can create the character to be you or be who you want to be. Or you could use your fears or create fears. And your first outfit might not be what you expect; it might not be what you want. Over time you'll develop it. My first outfit was pretty bad, but over time it got better through trial and error and help from the community.

SPM: *And where should people look for the elements of their outfits?*

LOCKE: Thrift stores in general. Target, Kohls. In our area, Savers. Marshalls, TJ Maxx, because they have good sales.

EVA: You can find good clothes there and easily steampunk it.

ADIE: I decorate my outfits with trinkets, because my character collects taxidermied animals, for instance. Gift shops, magic stores, thrift stores, antique stores, because they have mink shawls.

SPM: *And how much modification do you use to create your looks?*

LOCKE: For me in particular, I modify everything I get. I look at the original form and I see what it can be. The pants I use currently are plain brown pants that I striped using tape.

SPM: *Oh, I had no idea you did that! That's so cool.*

LOCKE: My boots are modded. I took two pairs of boots and put them together. My vest is an old jacket that has been turned into a vest. Everything I wear is modded, except for the red shirt I wear.

SPM: *So you suggest people change things up?*

LOCKE: Yeah, our group is a good example. Each of us actually had the same exact pair of pants at one point, but each of ours was a bit different from modification. We're gonna start helping each other out and making things together as a group.

SPM: *Are there any specific shops online that you would suggest for people to buy accessories?*

LOCKE: Etsy, because it supports artists. Gentleman's Emporium. Steampunk Couture... the guys' clothes there are really nice. Some of these sites are expensive, so keep that in mind. Our group might put some of our thrifty items up on the site, to help people out. And we would have information about how to mod them. And don't ever look up "steampunk" on ebay. You will be charged for the name. Look up "clock mechanisms" if you're looking for clockwork.

SPM: *Now, I wanted to talk about shoes. A couple of us here are shorties, and I noticed that you (Locke and Adie) wear boots most of the time, with heels. How do you survive in those? There's a lot of walking involved at conventions!*

LOCKE: They're broken in. And some people wear flat boots. But you need gel inserts, and if you get the right ones, they'll give you an extra inch too. The more you wear the shoes, the better.

SPM: *Alright. Eyeliner. Give me your tips, because you two (Locke and Adie) seem to be experts with it and I can't ever seem to get mine right. How do you do it?*

EVA: I don't know how they do it, but every time I put eyeliner on, it won't stay on.

LOCKE: Trade secret.

ADIE: I always put powder foundation on first.

LOCKE: I just put it on and then smudge it slightly. I put it on the inside of the eyelid and on the top of the eyelid, close my eyes and move it back and forth again, and smudge it.

ADIE: I use the cheap pencil kind from the dollar store.

SPM: *How do you feel that the steampunk life has affected you in your real life? Does it carry over?*

LOCKE: Actually yes. It's helped me feel freer and I enjoy myself more. I can use making props as a relaxation technique. It helped my style too. I changed my style to fit more to this lifestyle and it feels more natural. And I've made some of the greatest friends from this. So many people are so helpful and so kind. The steampunk community as a whole actually helps out so much. I know I can go to these people if I need.

ADIE: It helps motivate you to do other things too.

EVA: I've learned to sew. Before, I was wearing a corset from Hot Topic and a plaid skirt and goggles. But I wanted to get something better, so I researched about vests, bought two types of material. My sister-in-law helped me learn how to sew. She helped me make the vest, but I made most of it. Now I know how to use patterns and I know how to make a lot of things. Once I learned how to do it, it was so much fun and I just said, "move over, I want to sew!"

SPM: *Has this helped any of you be more outgoing?*

LOCKE: Adie here, before steampunk, she was that wallflower going, "hiiiiiii" but now she talks at panels, talks to people.

EVA: At my first convention, I wouldn't talk to people really unless my friends did, but now I can walk up to people even if I don't know them.

SPM: *Has it ever helped you get a job or make money?*

LOCKE: Some people make money off what they do, yeah.

ADIE: I started out with painting cheap squirt guns just to get my painting skills up. I sold them for very cheap because they were for people who are starting out and they could mod it more if they wanted. I also did simple medals and pins and such. They didn't sell as well as guns though.

SPM: *And now you're making those little vials you showed me earlier, the labeled ones with different liquids inside. Very cool. People can sell art or things they've made at conventions, right?*

ADIE: Renting a table at a convention can be a reasonable price or on the expensive side, it depends on how big the con is and various other factors. If you're selling art, though, you have to have an artist portfolio. That's something to keep in mind. Have photos of your work, samples of your art.

I ask what conventions are the most important to attend, and Locke is quick to tell me that all of them have different benefits and that there's no accurate way to rate their importance. He is conscientious of the hard work that has been put into the events and, indeed, is incredibly careful about the feelings of everyone he mentions during our interview. His politeness seems to be a throwback to an earlier time of manners, and even when he's silly, there's a sense of overarching goodwill.

During our day in Salem, we visit one of the very thrift stores where a steampunk could find vests, hats, and jackets to fill his or her heart's desire. Adie and I peer at various items and Locke and I find ourselves drawn to the same vintage military jacket. We discuss history and fashion and the second World War, and somewhere along the way I realize just how far steampunk reaches into every aspect of life. Whether someone is looking to delve into history or just wants to become more creative, there is a facet of steampunk that will benefit them.

As our visit comes to an end, I feel disappointed to say goodbye to this colorful collection of souls. There's an energy that surrounds them, a certain charisma that gains the attention of people wandering by. Anyone who doubts the power of science fiction has obviously never spent time in the company of people like the Vagabonds. Steampunk is far more than goggles or modified boots... It's a celebration of imagination, creativity and plain old good manners, no matter what time period (or dimension) you might find yourself in.

SEW YOURSELF A POUCH

By E.M. Johnson

This covered pouch allows you to show a bit of personality with a beautiful cover and classy embellishments. If you don't already have some spare buttons or fabric scraps handy, you can visit your local craft store, antique shop, or secondhand shop to see what kinds of interesting fabric, buttons, and accessories you can use for this and future pouches.

Adding a touch of sophistication to any attire, the pouch makes a useful and incredibly versatile companion for everything from travelling to daily goings-on.

Once you know the basics of making a pouch, the possibilities are endless. You can embellish it in different ways, experiment with different materials, and make it in a limitless number of sizes.

Materials:
- Fabric for interior of pouch
- Fabric for exterior of pouch
- Decorative/embroidered fabric for pouch cover
- Interior fabric for pouch cover
- Small piece of elastic
- 2 metal studs
- Button
- Thread

Tools:
- Pins
- Needle
- Sewing Device (Optional)
- Iron
- Scissors
- Fabric Pencil/Chalk (Optional)

Closures: Part 1

Gather your button, elastic cord, and all of the pieces from *closure fabric A* and *closure fabric B*.

To measure the length needed for the elastic closure of your pouch, take the button you are using and make a small loop with your elastic that is just big enough for your button to fit through.

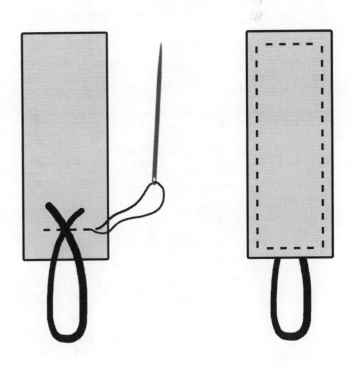

After measuring the correct size for the loop, stitch it to the "bad" side (the side you don't want showing) of a single piece of *closure fabric A*.

Place the second piece from *closure fabric A* on top ("good" side facing you) and sew both pieces together.

Do the same for *closure fabric B*, sewing both pieces together.

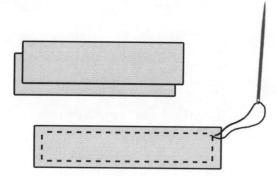

Pouch Cover

Take both pieces for the pouch cover and position one piece "good" side up, placing the finished product from *closure fabric A* as shown.

Place the second piece of fabric "good" side down (so both "good" sides of fabric are facing each other) and stitch the bottom and two sides together, leaving the top unstitched.

Turn the piece right side out after it is finished, so both "good" sides and the closure are showing. Iron until flat and edges are crisp. Stitch bottom and sides along the edge.

Pouch Body

Take both pieces of exterior fabric, placing the "good" sides together. Pin the edges of the fabric and stitch both pieces together, leaving the top portion open.

Repeat the same action with interior fabric, leaving both the top and a portion of the bottom unstitched as shown. (I recommend a 3.5"–4" opening at the bottom.)

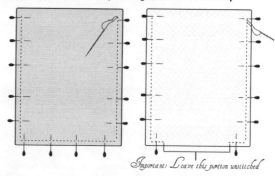

Important: Leave this portion unstitched

Remove the pins and flip your exterior fabric right side out, ironing until it is flat and edges are crisp.

Remove the pins, but do not turn the interior fabric right side out.

Retrieve the completed pouch cover and pin it to a single side of the pouch at the top (interior side facing out), and stitch it to the exterior fabric as shown.

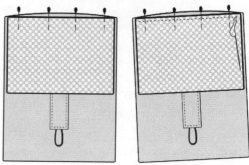

Proceed to tuck the exterior portion of the pouch (cover included) inside the interior fabric.

Make sure that the edges of the interior exterior fabric line up together, and that the side seams also match.

After you have lined up the fabric, stitch around the top so that the tops of the interior and exterior fabric are sewn together.

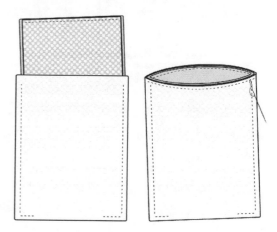

Flip pouch right side out through the opening at the bottom of the exterior fabric. (**Note:** This is one of the most difficult parts of the process, but take it slow and don't pull too hard, and it will be alright.)

Tuck the interior fabric into the pouch to create the lining of the pouch.

Closures: Part 2

Close the cover so it lays flat against the front of your pouch. Taking your fabric chalk/pencil (even most normal pens or markers will work), mark where you would like your piece from *closure fabric B* to be, along with your button.

When your pouch has been turned right side out, iron out any creases and stitch the opening at the bottom of the interior fabric closed.

Following manufacturer instructions, secure closure B to the front of the pouch using two metal studs. (Be careful to *only* put the studs through the front part of the pouch, not the back as well.) If you don't have studs or don't like the look of them, you can also stitch closure B to the front of the pouch. Sew button to proper position as well.

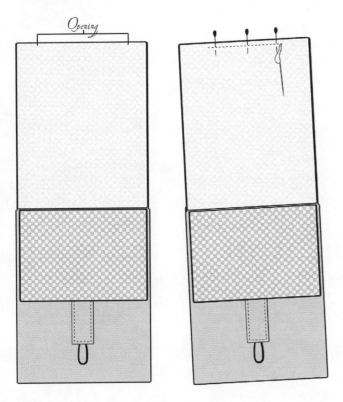

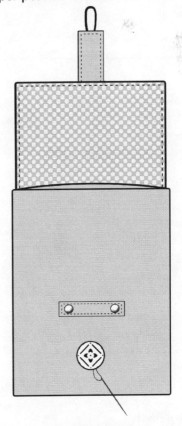

Making sure that fabric is properly aligned, stitch along the top of the pouch. (If you are making a smaller pouch, you might find it easier to do this part by hand if you have been using a machine thus far.)

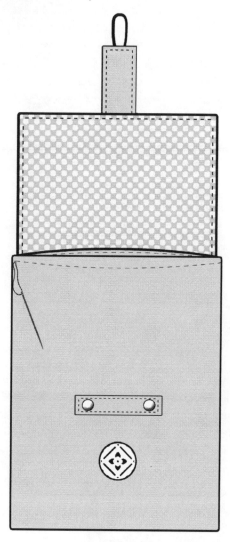

Belt/Purse Loops (Optional)

If you would like the option to hang this pouch from a belt or from straps, take the four optional pieces included with the pouch pattern and divide them into pairs. Stitch them together the same way as *closure B*. (**Note:** You can adjust the size/length of the loops to accommodate different belt or strap sizes.)

Stitch the loops to the back of the pouch, making sure they are neither too close nor too far apart, or your pouch won't sit properly on a belt/straps.

Variations on the Theme

Now that you know the basics of making a pouch, consider some of the following embellishments and alternatives to get the most out of it.

- If you want to make your pouch thicker/more sturdy, you can add interfacing to your fabric. I recommend iron-on interfacing, as it is one of the easiest interfacings to use with this project.
- Get creative with closures. Look for old door locks, keys, and other interesting objects to use as closures for your pouch.
- In this project you only use two metal studs, but you can certainly use more if you like the look, especially if you do this project in leather or suede instead of fabric.
- Think outside the box (or, a boxy cover at least) by using different shapes and patterns for your pouch's cover. You can use a rounded cover, a cover edged in lace, or even a cover cut into a creative shape or object (felt works really well for this kind of cover).
- You can also do without the cover altogether, making for a project you can create in a relatively short period of time.

Enlarge this pattern by 50% for correct sizes

Exterior Fabric
6" x 8"
2 Pieces

Interior Fabric
2 Pieces
6.25" x 8.25"

Cover Fabric
5.75" x 4"
2 Pieces

closure fabric A
1.25" x 2.5"
2 Pieces

closure fabric B
1" x 2.5"
2 Pieces

strap/belt attachment (optional)
1.75" x 3"
2 Pieces

SteamPunk Magazine #9   53

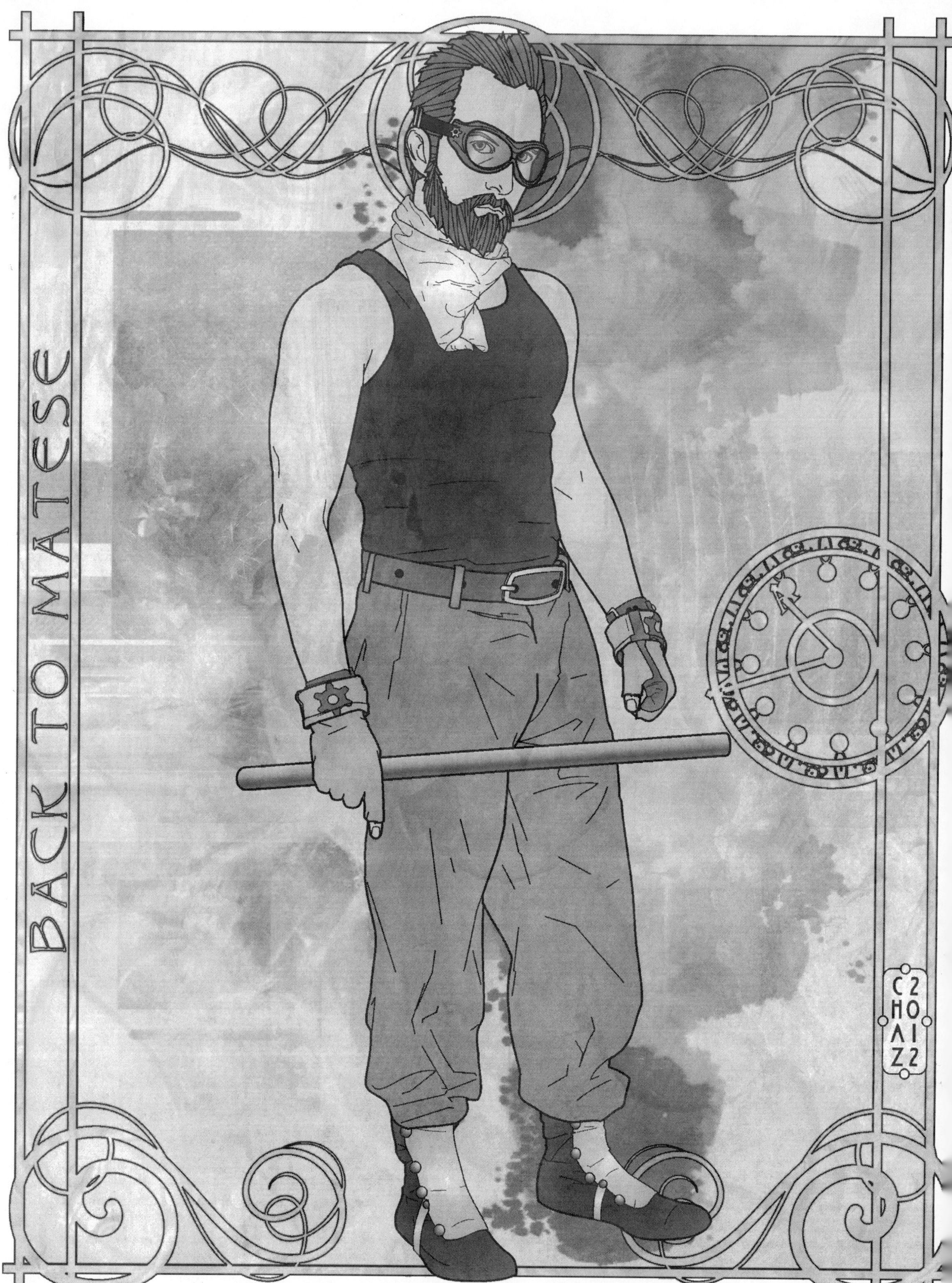
BACK TO MATESE

Back to Matese

Written and translated from Italian by reginazabo
Illustration by Chaz Kemp

Winter is coming in City 1, and the neige is falling thicker than ever over Richmond, settling on the ruins of Victorian villas that overlook the Thames and on the wide enclosed park that was once the Royal Family's hunting estate—assuming anyone still remembers.

Lumps of powdery snow fall down from the roof of the conservatory. During the summer, this sophisticated structure of glass and cast iron has absorbed the heat needed to feed the tubers and the timid coal leaves that sprout from the vases and boxes which—filled with a mixture of compost and clay—are spread everywhere: they are strewn across the chipped tesserae of the floor mosaics, hanging from the trimmed metal beams, clustered on the Art Nouveau wicker chairs with their cracked coats of varnish, and there are even some on the workbench and on the massive iron and brass calculator placed here back before the apocalypse, when the hothouse was turned into a laboratory by the will of a mad scientist.

A lot of stories still circulate about him, but few remember what happened before; as far as the Runts know, the old owner of the house might have dropped dead in one of his esoteric experiments—perhaps due to the green, glowing substance they found on the bottom of a glass jar. Judging from the unrelenting whistle of the radon detector, that guy had no scruples about collecting radioactive materials of any kind.

When they took away the glass of luminescent jam, however, the detector stopped whining and the Runts thought well of availing themselves of the opportunity offered by the conservatory, which had remained intact after the explosions and gave some hope of survival to their vegetables—their only sustenance besides the few rabbits left in Richmond Park.

After the conservatory, it was time to think about the house, which was fixed up some way or another and had immediately a couple dozen people squatting in it—or a bit more, counting the travelers and temporary guests who had to come down from their trees during the winter, the forest being even more impassable than usual when the branches froze.

Last night, they all gathered among the wooden columns and marble statues of the large smoking room to celebrate the arrival of their wandering friends and to cheer up before the big chill. Six months locked inside—the Runts are more used to it than the Old Ones, that's for sure, but when they're finished preserving produce, telling tales, and fixing things, they become just as anxious for the spring as the rest of them.

Disturbed by the insistent whistle, Georgie wakes up dulled. At the party he had gone hard at it, and with all that brandy mixed with the mushrooms, he must have gotten a bit carried away. He'd been with Mary upstairs, then with Nebo behind the chicory tank. Their skin, their tongues, the tang of sex in their sweat despite the cold draughts of wind—after that Georgie can't remember anything else, and most of all he can't explain how he ended up sleeping in the most uncomfortable place in the whole conservatory: the weird square cement platform, perhaps a pedestal built by the mad scientist to support who knows what kind of sacrificial rituals.

The whistle is getting deafening, and Georgie sets aside his drunken amazement to identify the source of the noise and suppress it. Lying on his belly, he tries to get up, but his neck hits something hard, which resonates with a metallic clang, like an empty tube. Only then does the young man realize that he is imprisoned under a strange contrivance of wood and brass—a vehicle it would seem, but for the lack of wheels.

Very slowly, because his head cannot handle any further shocks, Georgie crawls out of his cage and flops down on the floor. Then he looks around, exasperated, for the source of the alarm. On the workbench beside the door, the radon detector is going off like mad—it whistles, flashes, and hops on the stained marble bench. Georgie pushes a button and the cacophony fades away. The radiation must be damned close: it would be useless to let the thing's battery run down. The radiation—how did it get inside here? Maybe another bomb blew up?

Lost in thought, Georgie turns toward the middle of the room and gasps. The strange device had already slipped his mind. It was really a bloody daft idea to place it over him while he was sleeping. Who could ever think of making such an effort? And where did they…

In the centre of the machine there's a seat, or rather an armchair lined with leather and velvet that must have been red once, before a thick layer of dust tinged it in dark grey. On the chair, there is a letter. Georgie picks it up, annoyed, expecting to identify the authors of this trick, then he turns white and steps back—in front of the armchair, fixed at both sides of a console covered with lights and dials, there are two quartz levers with a green, glowing core.

With the letter crumpled in his fist, Georgie strides to the kitchen. "Hey, guys! Who's the dumbass who placed that clunker over me while I was sleeping? And yes, the spare parts can be useful, but did you see those levers? Couldn't you check before that nothing was contaminated? Radioactivity is so strong inside there that the detector almost shattered my eardrums!"

The kitchen is dirty and nearly deserted. Comatose bodies abandoned on couches, empty bottles, weak tea and tasteless porridge. That brandy has made victims of the whole house.

Mary, the only one who still appears vaguely sober, glares at Georgie and rests her forehead on her hand. "Stop yelling—my head is killing me. Is there some yarrow left in the hothouse? I want to make some tea."

"Whatever there was in the hothouse, that radioactive clunker has fucked it all up…"

"Eh? What clunker are you talking about?"

"I must have fallen asleep on the cement pedestal, and this morning I woke up to find this old crooked and wheelless car over me, with the detector howling as though a new nuclear war had broken out. And all the while, I had to struggle with that thing keeping me prisoner between its brass bars… You really don't know anything about it?"

"How do you expect me to know? I stayed where you left me last night. In your bed, until five minutes ago. What about you? What were you doing in the hothouse?"

Georgie pours himself a cup of coffee from the chipped porcelain pot, a graceful remnant of a prenuclear age, and sits in front of the young woman. "I found this on the seat. Look, it's addressed to someone: 'To my dear friend the Writer…' Can you read the name?"

"Uhm—Lales, perhaps? Malcs?… I don't know any writers anyway. Do you?"

"Stop kidding, 'course I don't. Let's see if we can figure out something from what's written inside."

Gallo Matese, Italy
12 April '77

My dearest friend,

From the moment we last saw each other few minutes may have passed for you, but to me it was days, and such thrilling ones as to change not only my life, but, I hope, the entire human race's, so that I am ready to bet that, should this Machine travel again to the future, all the way to year 802,701, it would find neither the Elois, those candid, harmless creatures, nor the Morlocks, their ghastly persecutors, but rather an endless Garden of Eden, where no living being would inflict domination and exploitation on their fellows or on other creatures.

That evening of no more than six days ago, which was but yesterday to you—when I appeared, ragged and hungry, during the habitual Thursday night supper and threw myself on the roast beef as the rudest of primitives—that evening seems so far to me.

How many hopes I had harboured towards my travels into the future—space explorations, peace, freedom from every slavery! Instead, alas, the human race had been replaced by other creatures, and all because of the hubris of the rich and powerful, who had pushed underground the industries and the working class until the day when the latter, forced to live with no sunlight, had slowly learned to do without it, evolving through the centuries into a progeny of rapacious, half-blind beings.

The sense of dejection I felt in the hours that followed my journey, reflecting on the sad fate of Humanity, was such that—in spite of my exhaustion and maybe also due to the meat I had eaten after days of strict frugivorous diet—I could not sleep a wink for the whole night, and in the morning I hardly managed to get up from my bed. If a bar of the Time Machine had not been damaged by the Morlocks and had not absolutely needed replacement, I would have lied there brooding over my disgraces until the end of my days. It did not go that way, however, and, luckily, in my closet I did not find anything that suited my purposes, hence I was forced to venture out to get a piece of metal with a length and a width suitable to replace the bent rail.

Wandering hypnotized along the streets of London, I found myself in Thames Street and mechanically entered Iremonger's shop, probably expecting that the poor old man would get out of his grave just to advise me about what metal alloy could best fulfill my needs. Mr. William Iremonger had been long dead, though, and his son, who was as into hardware as I am into embroidery, was out repairing something. It was a stroke of luck, because the shop had been entrusted to a clerk whom, before hearing his accent, I had already identified as an Italian from the thickness of his dark, curly beard and from his Sicilian cap. In fact, he provided me with many precious recommendations not only regarding the piece I needed to acquire (which I eventually found flawless), but also to overcome my state of desperate dejection.

The clerk turned the brass bar in his hands with a smug look in his eyes, as though he had guessed its purpose. Then, having established that among the wares in the shop there was nothing comparable, neither in size nor in shape, he led me to a warehouse on the back of the house, where the late Iremonger kept scrap parts of all sorts: burners for hot air balloons, mechanical clocks, astrolabes, and steam turbines.

"Here you might find what you need; you just have to rummage around a bit," he said, with his strangely amused gaze wandering on the tall heaps of scrap iron and gears that towered over our heads.

As I started to search that junkyard, probably out of trust for the sharp sight of the old shop owner, the Italian added: "If it is not a tactless question, sir, may I ask what it is meant for?"

As I have already mentioned, I was worn out by a sleepless night, and the fatigue mixed with the excitement of my travels must have distracted me, since I did not grasp the obvious technical sense of that question—the clerk having asked for enlightenment into the usage of that part within the larger structure where it was to be inserted—and I thought instead that he wished an answer as to the final goal of my mission, that he wanted to know why I wanted to fix my Machine and where—or better, when—I intended to go once I had repaired it.

"It is meant to save a candid and helpless maiden from the abuses of ruthless and bloodthirsty monsters," I replied without going into details.

"You need something else, then, if you wish those abuses to stop forever."

Thinking back, I cannot say what he had imagined, and if in these extraordinary days I have the chance to, I will perhaps ask him, but in that moment a shiver went through my spine, in a way that today I would call superstitious, and a belief arose in my brain—that the man knew, that he read my mind, and that he had guessed everything.

Perhaps the vibrations of my journey back home had shaken my head like some bland electroshock, perhaps my unconscious was able to choose better than myself with whom I should be allied—I crumbled. "So you doubt that with me Weena would be safe, even if I went back to rescue her!" I whimpered. "I would be forced to stay together with her, or to take her to a world where no one would be willing to accept her… what am I thinking?… to even approach her!"

"See, sir, the problem lies in this very division: if everybody could live with everybody else, if each person had what they needed and gave according to their possibilities, if all had equal rights and equal dignity, these abuses would stop. But sometimes I doubt that humanity will ever see the dawn of a fairer tomorrow…"

Moved by these words, and convinced that my supposition had been confirmed and that he knew something about my experiments and my wanderings in the year 802,701, I told him everything: how I had arrived in a place where civilization was unknown and how human beings had divided into two different races, how the first, the offspring of the most advantaged, had the attention of kindergarten children, while the others, the heirs of the working class, had turned into a progeny of wild beasts who nevertheless dominated industry and technology in a dark, malodorous, and impenetrable underground. I told him about Weena, and the Palace of Green Porcelain, and the night when the Morlocks kidnapped my friend, taking us by surprise in the forest. At the end of my tale, I had to hold back a sob, a sign of the exuberant emotionality that had caught my soul taking advantage of my state of deep exhaustion.

"Going back to save her will be useless," the Italian decided eventually. Amazed by such determination, I did not notice his impertinence, whereas in any other moment I would have hardly stomached an inferior's direct ways. "And I doubt there is much you can do in that world," he went on, "for, in order to change things, what we would need is a time machine."

That he read my thoughts seemed now apparent to me, and, putting aside my last reservations, I cried: "But that is what I was talking about! Time travel!"

The Italian looked me in the eyes, finally astonished, then started gazing at the junk heap on his left side, as to indulge in a moment of reflection. At last, having pulled out a brass bar which was only slightly longer than mine, as though he had always known it was there and had waited to make a decision before he extracted it, he offered it to me and said: "I really wish your story were true, sir, 'cause there was a moment, in my country, when everything could have joined together to change history. But only a handful of brave people realized it—just a bit older than boys, badly equipped and, what's more, betrayed by a vile informer."

The threatening look that thick beard conferred to the man's face was abruptly deleted by the softness of his broken voice.

"You were there... " I concluded. I did not find it difficult to imagine him, with those fierce, dark eyes, involved in a rebel plot.

His face darkened. "It was that betrayal that caused so many sorrows to your friend."

"Excuse me, but how can you think that a failed uprising in such a remote land could have caused sorrows in the far future?"

"Society is divided into two permanent castes, into two races, and, if your story is true, it is the proof that this division will produce terrifying mutations," he answered as if reciting a passage from a book he had learned by heart. "There is no natural law whereby evolution should inevitably give rise to liberty, or to the division of society, and only through the rebellion of subjugated classes can we hope in a better future!"

My dazed stare was read as an invitation to go on. "If what you are saying were true..."

His disbelief annoyed me. "Of course it is!" I burst out. "Do you perhaps think that what I told you was a children's story, some fib taken from a penny dreadful? One day, if you wish, I will expound the scientific foundations to you."

He seemed to ignore me. "If what you are saying were true," he continued brightening, "one could go back to the past to take advantage of the most favourable moment to change the history of Humanity!"

"Yes, sure, perhaps some day we can talk about it over a cup of tea," I cut short. I am known as a supporter of philanthropy, and the destiny of the human beings is close to my heart, but that day, shocked and sorrowful as I was, what mattered most to me was getting rid of the guilt I felt for Weena.

"But don't you want to avoid that your friend is torn to pieces? Don't you want to stop Humanity from falling into an irreparable degeneration, until it splits into two races of victims and persecutors?"

"Of course, but..."

"If this is what you want, then believe me: there will hardly be a new chance to give rise to a society where everyone is free and may live according to the dictates of love and solidarity." He seemed to be addressing a crowd, and as I watched him declaring his speech with his chest thrown out, raising his arm with a proud look in the eyes in that dusty, rusty little shop, I had to keep myself from laughing.

"And how do you think you could make a change, if I may ask?"

"If at the time we'd had the resources we have access to today, even the worst traitor could not have upset our plan, because the people would have sided with us—all the people from those villages, from that region, and maybe even from further beyond!"

"Right, perhaps even the whole nation would have sided with you," I replied. "But how many people are there in the kingdom? Some tens of millions, am I correct? What about the others? How do you think you can influence the life of the whole human race?"

"When the oppressed classes will see that their brothers live better without a government than by submitting to the coppers' and masters' hubris, you will see freedom spread like wildfire!"

"And when did this failed uprising take place?"

"In April '77, on a mountain not far from Naples, a city which at the time was, I am not afraid to say it, a nerve centre of the Internationalist movement. The band met in San Lupo, a village that, as you can imagine from its name, meaning 'Saint Wolf,' is perched on a wild and impassable land. Here, local peasants had just witnessed the succession of two monarchies without any real change being brought by the new one, which had allegedly been more enlightened but only made things worse.

"There was everything we needed: popular discontent, a new regime weakened by a decade long annexation war, the ideas of anarchism blowing in the wind to announce a better future… What was missing was suitable weapons, and an effective means to quickly persuade the peasants. But now we have all this…"

"And you wish to go back in time in order to carry out that endeavour with new means," I inferred, starting to nurture a certain scientific curiosity towards that story.

"Believe me, it would be the only way to save humanity," he answered.

"But supposing I agree to let you travel aboard my Machine, don't think you can load it with a whole arsenal of weapons!"

"Oh, that's no problem: what we need is just one gun… and one person."

"A person, and no bombs?" I asked in disbelief: how could he imagine to change the outcome of a rebellion with just one gun and one more man?

"Trust me," he replied, and I decided that I may give him a hand, if only to witness a rebellion with my own eyes; some days of rest before I went back to rescue Weena would not change much, and a little anarchic uprising in a remote land belonging to a small, nondescript state would certainly not cause great damages, I told myself. I was right, although I did not know to what extent, and I was so in a sense that I had not even considered. I gave him my business card and arranged to meet him at my home at tea time.

I spent the two remaining hours fitting the new bar and adding two additional seats in the Time Machine (fortunately, Mrs. Watchett remembered the armchairs I had bought at the Victoria Theatre auction, immediately before it had been turned into a movie theatre).

With the piece found by Malatesta (for this is how the Italian told me he was called), mending the Machine was child's play, and I even had some time left to prepare a knapsack with some rope, matches, and tools I could need in emergency, like a Swiss Army knife and a revolver. Considering that I was headed to the milder climates of the Mediterranean, I did not think, unfortunately, of taking with me a woollen cape or a raincoat, and luckily I put on my hiking boots, for if I had left with my slippers on as I had done in my first time travel, I would have certainly suffered from the cold of these snowy peaks.

When my luggage was finally packed, Mrs. Watchett came to inform me that tea was served: I had begged her to lay the table for three and to cook dishes that were substantial but light and easy to digest, so that we could be ready for our journey, but if I had known the identity of the third passenger, I would have probably asked her to use the finest service and the silverware, or, in fact, to lay the table just for two.

The tea was almost lukewarm, and I was availing myself of that wait by indulging in a nap, when the bell finally rang, nearly an hour later, waking me up with a start. Stuck in a nightmare in which the Morlocks pulled me into a dark pit, I was still recovering my composure when my housekeeper went to open the door and Malatesta came into the dining room, accompanied by a pretty and elegant lady and holding two large bags.

"A woman, Mr. Malatesta!" I burst out. After my mishaps with Weena, I was no longer willing to risk danger with a helpless creature to be protected. "If you really intend to go back to fight, you should leave your paramour at home."

The anarchist must have been in good spirits, for he burst out laughing and replied with a slight bow: "This is not my paramour, sir." The woman grinned mockingly. "And the lady you have before your eyes is not just another lady," Malatesta went on. "I am honoured to introduce you to Her Majesty Maria Sophie Amalie of Wittelsbach, Queen of the Kingdom of the Two Sicilies and consort of King Francis II of Bourbon."

Despite suspecting that the scene was just the continuation of the dream with the Morlocks, I jumped on my feet all the same to kiss the great lady's hand, and I mistakenly ascribed the sense of coldness I felt as I touched her fingers to my torpor or to the dream I had fallen into.

"As you may have guessed, I don't have a good opinion of aristocrats and hierarchies," the Italian specified, "but fortunately Sophie doesn't think much of etiquette and, to tell the truth, lately anarchists have definitely conquered her," he added, winking at the noblewoman, who turned to the other side to hide a burst of laughter.

"All right, Malatesta, I understand: a member of a royal family can always be useful, even if I find it dangerous—to say the least—to drag a woman deep into an insurrection… but I don't understand how you want to take advantage from this kidnapping: I thought that in '77 the Bourbons did not even reign anymore in Italy."

The queen intervened in our discourse, not upset in the least, and her voice was vaguely hoarse, as though it was reaching my ears from a telephone receiver. "Nobody has been kidnapped here, Mister —. I am here of my own free will, and my mission is to free Southern Italy from every king and usurper."

Spoken by a queen, those words persuaded me once and for all that I was dreaming, but I looked at Malatesta with an imploring gaze to figure out what kind of situation I had gotten myself into. "Trust me," he repeated for the second time in few hours, and his tone of voice was halfway between complicity and irony. "Wouldn't you rather show us your Time Machine and explain how it works?"

As you will surely know, my dear friend, talking about science is one of my greatest delights, and describing my inventions fills me with such pride that I could deliver dissertations for days at a time if nobody were to stop me, but this time I was brief: if I was dreaming, wasting my breath seemed pointless, and if I was awake, I wanted to get to the bottom of that whole story to achieve clarity… or simply the time travel attracted me even more than cold theory. After a moment, I led my visitors to the laboratory and explained that, besides the three more generally known dimensions, there exists a fourth, which we are accustomed to name "time;" finally, I briefly mentioned all the details we had thoroughly discussed with you and my other guests not earlier than two weeks ago in the smoking-room of my own house.

Partly due to my brevity, partly due to my state of mental confusion, I expected the anarchist to barely understand what I said and the lady not to follow even the fundamental reasoning. Instead, as soon as I stopped talking she asked: "So in practice, this vehicle can travel in the other three dimensions too?" The hoarseness of her voice was mixed with a French "r" and a strong German accent.

"Theoretically, it can," I answered, amazed by such perspicacity in a woman, even if so young and attractive, "but so far I have never tried."

"And how do you set the coordinates?"

I pointed at the controls on the console, and, as if she had been doing it her whole life, she quickly turned the knob until the hands of the quadrant, which I had never moved until that day, indicated a position that the lady must have memorized before and which I found plausible: 41 degrees North, 14 degrees East, at an altitude of thousand metres over the sea level. Then, without saying a word, she sat on the theatre armchair and glanced at Malatesta, inviting him to sit in the passenger's place next to her.

Remembering that our tea was still waiting for us on the table, I ran back to the dining room to pack some sandwiches in my knapsack, which gave me the chance to grab my camera. It was in that moment that you arrived to visit me and found me in the smoking-room. But I could not make my companions wait any longer, and I was impatient to set forth on our journey, so I greeted you fleetingly, being sure—as I told you—that I would see you in half an hour. If my calculations are not mistaken, the Time Machine will be back on time: I will not be there, and what follows should explain to you why.

After I greeted you, I ran back to the laboratory and immediately sat in the driver's seat. Having made sure that the passengers were suitably bound to their armchairs, I pulled back the quartz levers, expecting to see, as had happened before, the signs of time flowing, indicating a movement along the fourth dimension.

But this time, perhaps due to the additional spatial movement, the Machine was wrapped in a white blinding light, and a strong draught of wind nearly tore us from our chairs, forcing us to cling to the side bars of the vehicle. Then the machine started rising, apparently crashing the roof of the conservatory, and I felt relieved that I had repaired the part that had been damaged by the Morlocks before I started my journey.

I glanced at the time dial: we were already in '77 and were bound to reach the month of April in a few seconds. I slowed down our pace, hoping not to be engulfed within a mountain, and the view that took shape before our eyes stunned me: we were floating above a landscape of pure snowy peaks and mountain lakes, and under us forests and meadows spread as far as the eye could see, dotted here and there by Medieval villages whose grey brick houses were perched on the hills scattered in the valley. Seen out of time, it was an idyllic panorama, and only when we stopped did I understand, from the neglected look of the buildings and of their inhabitants, why that was the place where the revolution must begin. "What day exactly?" I asked my passengers.

"Sunday, the eighth of April," the lady promptly answered, not bothered in the least by the hair tufts the wind had pulled out of her chignon. A few moments later, I stopped the vehicle: it was the morning of the eighth of April, and we found ourselves on the shore of one of the high mountain lakes we had just observed from our atemporal vantage point. Behind us there was a peasant village, and from the path that led to the top we heard voices singing:

Is it the eternal suffering is not enough?
Stand on, brothers, everywhere, at the same time!
Defeat, kill them, curse the criminals!
Enlighten the dawn of a better life!

Presently a black and red flag appeared behind the ridge, and thereafter we sighted a group of hats decorated with cockades of the same colours. It was the Internationalists, Malatesta's friends, and they had reached the village to announce the advent of the social revolution!

With my companion's help, I hid the Time Machine behind a bush while the queen extracted a coarse shepherd's cloak from her bag and pinned to it a brooch identical from all points of view to the Internationalists' cockades.

When we arrived there, the village was in havoc. Past the dilapidated huts that flanked the village, along the more decent main street the shutters overhung by Neoclassic gables were bolted, and the piazza before the church and the Town Hall were crammed with a crowd of beggars who encircled the flame of a recently-lit bonfire. From the balcony in the front of the Town Hall, young men not older than twenty years with red scarves around their necks hurled large folders of yellowed papers into the fire. At each throw, the mob cheered.

We did not seem to be needed much, and perhaps my fellow traveller noticed some incredulity in my eyes, or in fact he read my mind once more. "A thrilling view, isn't it? If one witnessed only this scene, they could believe that the revolution had broken out. It's a pity that we were just twenty-six, and right now a good twelve thousand soldiers are being deployed in the surrounding villages, resolved to stop us at all costs from putting right history's mistakes."

"Twelve thousand soldiers? If you think that this time you can have the upper hand over such an army, you must have a really lethal device in that sack of yours."

"Ah, well, yes, lethal it is: though small and compact, it can flash a powerful beam of energy to any distance without the slightest dispersion. But the instructions to build it are easy to track down, in our time, and my secret weapon is not hidden in this sack. Persuasion, as subtly manipulated as it may be, will be much more effective than firepower…" The anarchist nodded at Maria Sophie, who was treading towards the centre of the square few steps in front of us, while the dumbfounded villagers turned to stare at her as if wondering in which picture they had already seen her face.

Perhaps spurred by the excitement of homecoming, the noblewoman was leaving us increasingly behind. I availed myself of her distance to express a doubt without offending her. "But don't you think," I whispered to the Italian, "that a queen, though with the best intentions, will hardly help your cause, which aims at knocking down the kings and the powerful?"

"Sophie was a queen for barely a year, from May '59 to early '61, after which she had to abandon her kingdom forever, and she never stopped missing it." I observed the dignified lady solemnly walking among the crowd, with a lofty bearing despite her raw peasant woollen cloak, and was struck by her smooth skin and shiny hair—if what my companion said was true, that woman had to be older than me, yet she was as radiant as any movie starlet. The anarchist went on: "In '77 we hadn't tracked her yet, and since we managed to convert her, there has been no opportunity to put her potential to good use."

Malatesta's words were vague, and I was going to ask him about the details of that "conversion" when I was distracted by the weird behaviour of the crowd, which opened up, gaping at us as we passed.

"Just as I had imagined," my companion said, observing the villagers' wide-open eyes, their malarial gaze, their cheeks hollowed by malnutrition, "hardship produces a depressing, and therefore counter-revolutionary effect, but in order to overcome this hindrance what we need is an icon, a symbol, provided that it does not become more important than the cause itself. Now will you pardon me? I'd better take care of what must be done." And he ran forward, to confer with the queen. He then made her stop a few metres from the Town Hall arcade and lingered awhile, adjusting the buttons on her back before he set forth directly, entering the large hall together with her. The masses closed around me, stopping me from following my fellow travellers inside the building.

Rather than elbowing my way through a horde of people whose language I did not understand and where I was clearly an alien element, I decided to climb to the top of the church stairs in order to watch the scene from above. On the other side of the bonfire, in the middle of the square, stood a stone with a cross where someone had tied the black and red flag that we had seen earlier waving over the marching insurgents' heads. Standing on the stone, upright and proud with his cockade pinned to his hat, was someone who I might have mistaken for Malatesta's younger brother, had I not been aware that I had travelled eighteen years back in time and that I was actually seeing that very anarchist, when he was twenty-three years old.

This man had a thick, dark beard too, though not as long as my companion's, and his eyes were burning embers like his. I did not understand what he said; I do not even think he was speaking in Italian, but I did make out the words *rivoluzione sociale*, and during the speech that followed, the bystanders repeatedly nodded in assent, muttering comments and cheering as he spoke. A peasant woman pushed her way through the crowd and examined his face, as if to gauge his sincerity. Then she talked, and the others reacted to her sentences with whispers of approval. The young man grinned mockingly and his answer was met by many people spitting on the ground or shrugging. Afterwards I was told that the woman had asked him to take care of the division of the land, and that to this request he had answered: "We've given you guns and axes; you've got knives. If you want it done, you'll do it, otherwise you are screwed!" And even if I had understood that the band wanted to inspire revolution without commanding it, I set aside any amazement I could have felt regarding its failure. Which uprising can be successful, I wondered, if its inspirers are not trusted by the masses? It was then that I started grasping Malatesta's discourses on the arms of persuasion.

After the exchange between the young anarchist and the peasant woman, I had the impression that the bystanders were dispersing, but in that moment a lanky, spectacled man in his thirties ran out of the Town Hall, conferred for a minute with the young speaker, and called up the others who were inside the building. Presently, a haughty and ever-imperturbable Maria Sophie came out of the main door, and once again the crowd opened up, letting her through with a diffuse murmur. Many people, especially the older ones, had recognized her by now, and the words 'a regina, "the queen," came to my ears from many directions. At that point the lady spoke, and silence fell on the square—they listened to her in a trance, as though a Madonna had just appeared before their eyes on the stone with the black and red flag.

I do not know what she said with her strange hoarse voice, but the last word she spoke, clearly articulating each syllable, was: *anarchia!* The exultant cries of the peasants, who until then had welcomed quite coldly the gift of liberty, moved me to the point that I started crying together with them, and exchanged hugs with people who I would have approached only with reluctance, had I met them in normal circumstances, in the meadows of Dorset and Cornwall.

Malatesta reached me outside the square. "Our task here is done: let's go refresh ourselves, and I will introduce you to the others!" He was beaming. We found the Internationalists—twenty-six in all, as my friend had told me in advance, not older than forty and not younger than nineteen—celebrating in the only tavern of the village.

As we could have expected, we were inundated with questions and had to explain the whys, the wherefores and most of all the whences of our journey, but eventually we were given a moment's respite, and I had the time to satisfy my curiosity about the only woman in the party.

That guerilla warfare and revolutions were not a prerogative of the male gender was very clear to me, and I remembered that in both the Paris Commune and in the insurrections that had shaken Europe during the whole century there were educated and brave women like Louise Michel or Vera Zasulič, not to mention the suffragettes, who terrorized the respectable British subjects of my time by claiming the right to vote, along with other alleged universal rights. But that a queen could be willing to go back to her adopted country to renounce all her powers and plunge into the midst of an uprising to champion the anarchist cause was incredible to me, and I could see no explanation.

"What is a queen," Malatesta told me, "but a machine governed by known and modifiable rules? One just needs to learn those rules in order to change them and to turn a foe into our most faithful ally." Noticing my astonishment, my companion reflected for a while, then added: "You surely know that for more than a century the royals and aristocrats have been replaced with automata so as to keep order!"

"No... I... had no idea," I admitted, wondering how royal families could have been replaced a hundred years before with automata that were so complex as to look real. Technology, I had to acknowledge, thinking about Leonardo da Vinci's machines or about the Song Dynasty's fireworks, had neither originated in my time, nor in my country.

"After the French Revolution, the great international financiers, the Rothschilds and their enlightened friends needed tranquillity to settle their affairs, their world. So they made a first attempt with Napoleon, but he slipped out of their hands. The first reliable results started to be achieved in '15, when the stable versions of their androids were installed in the thrones of all Europe. The Congress of Vienna was nothing else but this: while the machines boasted their highly advanced oratorial abilities at the Ballhausplatz Palace, bankers and Freemasons uncorked bottles of champagne together with engineers and inventors in the foyer at the ground floor.

"But of course these mechanical kings must be dethroned from time to time, even if they have proven ideal, to ingratiate themselves with the people, which is their primary function, after all. When the French and the Brits decided that Italy had to be unified, the obvious conclusion was that Maria Sophie and her royal consort should be deactivated. They would be switched on again every now and then: at a reception here, at an opera there, just to make some impression, just as with the rest of the impoverished nobility waiting to be taken apart for scrap."

"Now I understand why she has such a disquieting voice: I thought she had acute tracheitis. But how did you find her?"

"Ah, Sophie is a really extraordinary model for her age," Malatesta said, smiling. "Wittelsbach AG only produced two specimens of it: she, identifying code M50F14, and her sister Sissi, or better: M51S51. But actually these machines are too advanced: the producer supplied them with an emergency solar panel that reactivates them if there is a brownout. Think about the ladies who faint all of a sudden—Wittelsbach AG thought well to avoid this inconvenience, so these androids cannot be possibly switched off.

"For Empress Elizabeth this is not a problem, since her empire seems destined to last for centuries… as long as the social revolution can be prevented…"

"But Maria Sophie should have been kept in a closet…" I concluded.

"Right. At the end it was resolved to lock her up in a villa at the outskirts of Paris, and once she arrived there it wasn't hard to approach her, with the excuse of some maintenance to the electric system, and to change the rules of her operating system."

"So your delay in meeting me at my home…"

"The airship from Paris had had some problems at take off," Malatesta explained apologetically.

Thus, my dear friend, being satisfied with those answers and convinced more than ever that in order to advance human knowledge one must look in the most unthinkable corners, I spent the night in company of my new adventure mates, and I tried to exchange a few compliments even with the android, who was entertaining her bold fellows with the grace and coquetry of a great lady designed to seduce. When rewriting the commands that governed her valves and gears, the anarchists of our time had thought well to keep intact her deepest nature, the most powerful weapon she had.

Now that, thanks to the skills of Malatesta and his comrades, Maria Sophie has sided with them, the peasants of Southern Italy will finally place their hopes in this war of liberation from abuses, and the seeds of freedom will be sown in these remote lands.

We saw it that day in Letino, and the day after in the neighbouring village on the other side of the lake—where I find myself presently—and when the first Savoyard battalions were sighted, the troops from the South joined our side as soon as they saw the queen riding in front of the peasant army.

Our numbers increase by the day, and the authorities relinquish their command as soon as we approach. I have resolved to stay here to take part in this struggle until the end, because I believe that this is the ultimate battle to redeem humanity.

In order to send you this message, I needed to sacrifice the Time Machine, of course. I am doing it with good grace: if the effects of this decision prove fatal, you will already know, and might decide to go back in time to save me. But if, as I imagine, the promise of a better world will come true, this method of escaping my destiny and my sad era will be of no use to me.

I wish you well and greet you with most sincere affection.

Yours —

"The signature is undecipherable," Georgie says in the end. "We will never know who's written this letter and whom he wanted to send it to."

Mary throws a handful of white, dried florescence into the water boiling on the wood stove. "So that thing in the veranda is supposed to be a time machine?"

"It's clearly bullshit: assuming it works, how could anybody build a time machine in '95? Where would they find the spare parts? There is not one functioning industry left, and mines are all abandoned…"

The girl dips a blackened teaspoon into a jar filled with raw honey and moves it into the chipped cup, stirring her tea absent-mindedly for what Georgie perceives as an unending lapse of time. "Have you ever heard of this anarchic insurrection? Where was it, again? Italy? Do you remember where Italy is?"

"To the South, on the Mediterranean Sea. There is nothing down there now, in any case."

"It's there that the first explosions took place, right?" Mary asks, massaging her temples.

"How would I know? At that time I was little more than a toddler…" Georgie stands up and turns to his friend. "I'll try to figure out how to disassemble that whatsit: would you like a brass plate to stop that leak in the ceiling of your room?" ✺

This story was originally published in Italian by the Italian steampunk fiction journal Ruggine *(COLLANEDIRUGGINE.NOBLOGS. ORG) and appears here with permission from the author and translator. This story is not Creative-Commons licensed. Instead, "the authors humbly put this story at the disposal of those who, in good faith, might read, circulate, plagiarize, revise, and otherwise make use of them in the course of making the world a better place. Possession, reproduction, transmission, excerpting, introduction as evidence in court and all other applications by any corporation, government body, security organization, or similar party of evil intent are strictly prohibited and punishable under natural law."*

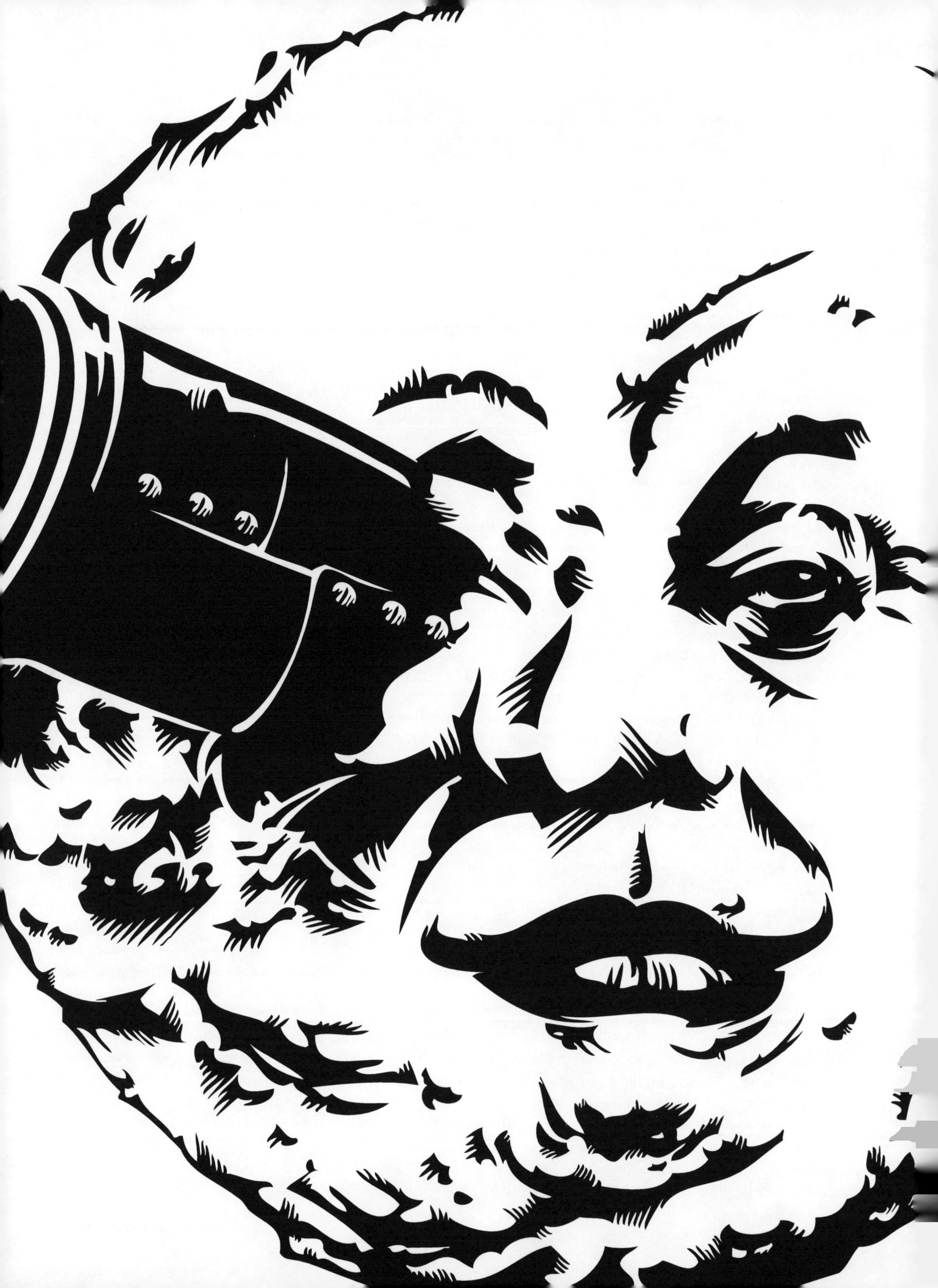

A Trip to the Moon

An Interview with Eric Larson, the Man Behind TeslaCon

By Kendra Saunders

Note from the editor: this interview was conducted in spring 2012, approximately six months before the convention in question. We are running it now because we find it a relevant and fascinating look inside the mind of a convention organizer.

I had the great honor of talking to Eric Larson about Tesla-Con, a vision he has created from the ground up. The amount of creativity that has gone into this convention is astounding. Eric details everything from tips for building your own event to how to become a performer at TeslaCon. Welcome to his mad, mad world.

STEAMPUNK MAGAZINE: I didn't have the honor of attending TeslaCon last year. What can you tell people like me who are brand new to your event?

ERIC LARSON: The convention is an experience, unlike any other. It was designed that way for a reason. So many conventions are based on the early 80s premise of cons. I grew up on Star Wars and know this very well… dealers, panels, and entertainment. I don't think that's bad really, it has worked well for over 30 years. The issue I have with the panels and such, [is that] even the dealers all start looking the same. I went into creating TeslaCon to be different from the very beginning. I vet all dealers, I help create or work with presenters on most of the panels, and we do different things from the very start.

We have a fashion show, and we now will have a full costume contest. However the feel will still be slightly different. The panels are created to reflect the year's theme. Since this year is about going to the moon, many of the panels will have a feel for that, or at least encompass something of it. We are also including Cthulhu and that entire story into our storyline and panels. I like to give the convention depth and a feel of realism.

We estimate about 98%–99% of people are in *full* costume/period dress. This is really different considering many also stay within character throughout the con. Well over 50% love to become their persona and make the weekend around it. Manners-deportment and etiquette are all observed. The funny thing is I have never really asked people to do this; it just seems to emanate from the attending members.

SPM: *See, and that sounds absolutely awesome to me. As a writer, it's always cool to be around people who accept that level of imagination, acting, and immersion into character. Now, two years ago was the debut of TeslaCon. How long did it take you to create this event?*

EL: We started in the early fall of 2009, by kicking off at WindyCon, the year they did a tribute to steampunk. This is where I met many of the folks who eventually would be helping me out. Each con takes about 14 months to create. Working 6–8 hours a day for 6 months and then 10–12 hours the latter half. What makes it different is the amount of work that goes into each item-from posters and wall art to passports and small items placed around the hotel itself. Anything and everything to create ambiance and decor of the age.

SPM: *Wow. How do you go about finding guests for TeslaCon?*

EL: Finding guests is really hard, mostly because I am picky. So many people write me asking to be a guest, but I never find that they have much Steampunk experience. Or they ask for so much money. It's merely a show for them, and I won't deal with that at all. I feel all my guests so far have been absolute gems. This year will be slightly different, but I think what I'm aiming for is going to be fun. I really consider the "ships" to be my guests also. They come to perform, but really do end up creating a feeling of immersion for the members that I could not get out of anyone else. I see who is popular at the moment, but also make choices on what they have done what they are doing and if I like their work.

It's gotten to a point where they find us, seriously. Our first year we did around 460 members total, that included staff. The second year we topped at 790 with staff, the third year we are at 700+ people for preregistration and that is of March 1st, 2012. It's growing because people tell me they're yearning for something new, something they have never experienced before. Many letters I receive are from people who live anywhere from 100 to 300 miles away. This last year we had someone from Alaska fly all the way in because he heard of the con. Others from the Seattle area and as far as Maine also make it in.

My job was and always will be to create a world where you can have a journey through my world, to experience Steampunk with your friends and others in a new unique way that is both for a group and individualized. I attend other cons, and also have surrogates helping me in other areas of the country. I hope to start having supporters overseas as well. We still reach many through Facebook and now Google+, but also Twitter and blogs. Many of our fans do the best amount of advertising for us. So many have written in the past saying that they want to attend based on what their friends wrote.

One important thing we did learn this last year was how many wish to share the experience with their families. We had a pickup of 4 to 1. For every guest last year, someone brought one to three more family members. One family out west brought 8! We do lots of one on one also—I enjoy that the most. Just meeting the fans at cons and explaining what we are.

SPM: *How should steampunk performers contact you if they're interested in participating at TeslaCon?*

EL: Simple—write to me on the website. There is a link. This year we have more people and the entire site will be redone as of April 1st, 2012. Contact me directly and you will receive response within 72 hours usually.

SPM: *How do you and your convention help build the careers of steampunks?*

EL: It depends what people want out of it. I have just learned that TeslaCon is considered a place where this happens, so I want to help as many people as I can. Such groups as The IAPS from Michigan, The Stafford Society in Milwaukee, The Vagabonds from Massachusetts, and many others all come to have fun. But the convention really does help them by exposing "their" brand of steampunk to the masses. What should be understood is everyone is picked for a reason. Jenni Hellum plays Kapt. Von Grelle, a wonderful character and persona. I sit down with them beforehand and explain the storyline of the year. We incorporate as much of their story into mine, but also make it unique for the convention. Her character will have much more to do with Doctor Proctocus (our villain this year), while others will be used in other supporting roles.

What I enjoy is helping these entertainers create a persona, but also learn how to share it with others. We actually have classes in this at the convention. The Ace of Spades group from Michigan ran a wonderfully detailed panel on this in year two. I am always looking for more entertainers, but also look for new and different acts as well. People who can add to the story but also bring their unique perspective and ideas to the table to make this event more than what they thought it could be.

It was funny, this last summer at a Chicago convention my good friend Montague Jacques Fromage told me over breakfast that I had a nickname. I laughed and asked what it was. "The Disney of Steampunk." He thought it was very fitting. I wasn't sure at first if I liked it, but after a while and with his reasoning it made sense. I create worlds for people to visit and give the performers and entertainers a chance to become part of the world. And by the same token, I give them a chance to stretch and make their performances new and engaging with varied audiences. So I guess I am like Walt to some degree… but I don't have a castle. Darn!

SPM: *We should build you a castle… maybe one on the moon? No, but really, I can see why people would hear about this and want to go right away or bring friends. I want to go now! Can it be tomorrow? Ha! So, what were some of the unexpected challenges of creating TeslaCon?*

EL: This question could fill a book. I will start by saying that I truly do love making the convention, and by making it I mean creating to the level I do. It is very rare that a person is given a chance to create and then "unleash" their creation as I have. [My creation] is watched, and also performed in and made real by people's imaginations and talents.

The challenges are very different from other conventions. Most worry about badge art or program books. I worry about how I will create an original hour's worth of immersive film with special effects and sound design and make it so the entire convention engages with it as I want them to. Herding 1000+ people into a room to get them to experience my vision is harder than it looks. We are starting earlier this year on creating robots and alien-style topography for the film. Using the Melies film as a springboard, but also creating new designs and having fun making it a TeslaCon branded world.

The members also are a large part of making it real. They seem to, without me pushing it on them, create a world of their own inside of mine. They embrace it and make it real to themselves. It's like giving everyone a crayon and a piece of paper with a theme, and then letting them draw what they want. It all seems to work in the end. So far I have been lucky.

SPM: *What were some of the major improvements you made to the event by the second year?*

EL: More staff for one thing! I have a very dedicated group of individuals that give their time and effort in creating my world of Steam. From Fixer, a dear friend who has worked over 15 conventions of mine, to all the new students I bring into the fold. This year Katherine Gonzales will be our volunteer coordinator. Along with the most famous Tea Lady in the world! There are many people that make this happen—but I also made adjustments to story and look. The second convention was far more immersive, as the third will be. If the first was 100% and then the second was 200% we are on track to create about a 320% improvement this year.

Because we sell out, the memberships move faster when entering. It takes only a moment to get through. No money is exchanged, but you simply check in and go. One feature we are doing the third year is *instant* sign up. Get your room for next year also! Since year four is the *big* year (The First Congress of Steam) we want it to go very smooth. So this will be the test run.

Another thing we added was surround sound. As you went through the halls, ship sounds and voices were heard around you. It gave the entire convention a more realistic feel. This is something we will continue to do as well.

SPM: *What are the top five most important things someone would need (capital? A magic wand? Eighteen trays of stale bread, ala Arisia?) to create their own steampunk convention?*

EL: Money is important, at least five thousand to start. Anything less and you will be just making do. It's sad really, because I think many people have good ideas, getting them off the ground is another matter.

Second would be a *great* negotiator. The hotel wants to make money, you need to make money... you both need to do well but must work together. This really is the main issue with most cons. If the hotel is stubborn or won't work with you, it ruins everything. So far we have had great luck with the hotels we have picked.

Staff. This does not mean all your brothers, sisters, and family. This means people given exact assignments with reasonable deadlines and outlines of what is expected of them. We all have problems and issues, but this is a big problem at cons. Some people try this with a bare minimum and you can't. It will ruin your health or more importantly your sanity.

Other things would be a clear vision of what you want and why. Decent dates for the con. Don't go up against a pre-existing huge convention... ever. Finally, panels that make sense. Have a vision of what you want to see and ask people what they want. Don't bother with all the same type of panels you see everywhere else. Try something new and encourage diversity in them.

SPM: *What are the five most important things an attendant needs to bring with them to your event?*

EL: One: a great sense of humor. No kidding. TeslaCon can be serious, but it is more fun than most people would think. People really enjoy themselves at this event.

Two: The semblance of a costume. Even if you can't afford a whole steampunk look, make an attempt. You will have more fun and people will help out—*really*. Khakis or black pants, plain shoes (no tennis shoes), and a simple shirt either white or tan. We are thinking about making clothes for people to use during the con, so that if they can't afford much more than basic "simple" they can at least have fun mixing

> Never let anyone tell you what you can and cannot believe in. Unlike many other fandoms that are told how to dress and what to wear because there is a certain convention of rules one must follow in costuming, TeslaCon breaks many of those old beliefs.

and matching. Most members find that TeslaCon attendees are very nice and do care about each other. That's what makes it such a rare con.

Three: appetite. I have the only convention that actually themes its food around the theme of the convention. We have our Executive Chef start preparing recipes long before the con and encourage him to try new things. This year Chef Matt will be doing some very fancy cooking with dry ice and such. The reason—we are on the moon! Who wants bland boring hotel food when you can have Orion eggs and puff asteroids! The other fun thing is we create the hotel's menu. And rename dishes in our vernacular.

Four: spending money. We are expanding the dealers room to be three times the size this year from last. Because of the date, this will be the one place everyone should come to buy their friends and family something steampunk for the holidays. We are also leaving the dealers room open to the public, in hopes to make converts. I have gone through over 139 dealer applications to pick what I hope will be the most diverse and fun dealers room at a North American convention. *All* dealers must sell at least 90% steam-clock-diesel ware. But we are encouraging people to pre-buy and then pick up at the convention. Maybe a new dress or that special Nerf conversion you have always wanted. This will be a great weekend for deals.

Five: finally and maybe most important… *imagination*. I take great pride in knowing that people want to return because it makes them truly happy. For the first time last year we introduced a dance called the "Choo Choo Train." It really is a gigantic conga line. When viewing it on youtube I heard something and saw something at the exact moment I could not see while there. Someone shouts at 1:08 into the film "This is Fantastic!" It hit me… everyone in that room had a huge smile on their face and was laughing and really was having a genuinely good time all being together. Never in my dreams could I see this coming, but it really is the essence of the con.

When people arrive I greet them, in person. I tell my staff to do the same thing. We help you into line, carry your bags to your rooms and if you need extra support, we are there for you. From the moment you arrive you will believe something is different. The sounds, the announcements, even the venue is decorated to the Nth degree. Bathroom signs look vintage, new maps are posted on walls—rooms have names based on the theme. Even the hotel room doors themselves are decorated in a new manner. Imagination drives me, and it drives the convention. I worked on *Star Wars* for three years creating art and design for the re-release back in 1997. I also have worked on Disney and Warner Bros. films.

What I have learned is to never let anyone tell you what you can and cannot believe in. Unlike many other fandoms that are told how to dress and what to wear because there is a certain convention of rules one must follow in costuming, TeslaCon breaks many of those old beliefs. I wanted a convention where fans could really take charge of their own imaginations and create the person or celebrity they wanted to be or to portray. I have strict rules on saying anything about costumes. If you don't like the way someone is dressed, fine. Keep the comments to yourself. Nobody should ever feel they are not welcome because of their design or applied artistic style. I love to see all types of costumes and period dress. From clockpunk to steam to diesel wear, it all fits in with who and what we are. I encourage people to be daring with design and with their personas. I want to see people make a spectacle of themselves (in the right way) and create with that a story that helps propel their design or character even more.

TeslaCon is about being creative, exploring your ideas and imagination. [Imagination] is one of the most precious gifts we have as human beings, and I love to promote it as much as possible.

SPM: *I interviewed the Vagabonds recently and they spoke very highly of your event. They specifically mentioned they liked the immersion aspect of it, and the story that went along with everything. What do you think it is about TeslaCon that keeps people coming back?*

EL: TeslaCon is above all things still just a convention at heart. I do many weird, wacky, fun, and extremely big themed events all in one weekend. People need to get away and have fun! I don't really know how I do it, or how it works. My staff would say about the same thing. I consider myself very lucky and fortunate to have so many people believing in this convention and what it is I do.

I usually try to say goodbye to most attendees when we are done, and I am always surrounded by fans. It is very humbling to know that you can touch so many people in such varied ways. I think two examples sum this up:

For two years the same young man has approached me and thanked me profusely for throwing the convention. He openly cried. He said, "You made it believable to me."

Then the woman who has never attended a steampunk convention before in her life. She curtsied and asked to speak with me in private. She explained that this last year was very rough on her, she had lost her job and her house. She decided not to give up. Well, she found a home and had gotten a new job, but still wasn't feeling great about things. One of her friends told her about the con and she decided to attend. She gave me a huge hug and grinned from ear to ear and told me that she had never felt so much a part of something as she had this past weekend. "I can't wait until next year," she said "This has given me so many ideas I can't wait to start."

PUNKING THE PAST:
THE POLITICS OF POSSIBILITY

by James H. Carrott

Do not be afraid of the past. If people tell you that it is irrevocable, do not believe them. The past, the present, and the future are but one moment. Time and space, succession and extension, are merely accidental conditions of thought. The imagination can transcend them.
—Oscar Wilde[1]

IT'S NO BIG STRETCH TO IMAGINE THAT OSCAR WILDE would appreciate the hell out of most of what steampunks do. He tweaked the nose of convention, transgressed with aplomb, and never let a fact get in the way of the truth. It's worth noting, however, that the word "steampunk" would have suggested something a bit more explicit to him. In Wilde's day, as a noun a "punk" was a prostitute and as a verb the word had begun to develop some pretty specific connotations related to gay sex. Heck, the *Wilde vs. Queensbury* trial even got, er, straight, to the subject when Queensbury's lawyers played their scandalous trump card, calling out the great aesthete's interest in punking punks (Wilde was then forced to drop his libel charges when the court decided that he had indeed been "posing as a Somdomite [sic]"). In sum, he would say it better than I ever could, but it seems quite safe to speculate that Oscar Wilde would have (off the stand; homophobes held the gavels) admonished us all to punk the past with a passion.

And he'd've be right to do so. History isn't sacred—historians are neither popes nor priests, and the past doesn't live behind bullet-proof glass. So why not punk the past? In contemporary parlance, "to punk" has taken on a more PG-rated image. After all, Aston Kutcher "Punk'd" 64 episodes worth of celebrities on broadcast television from 2003–2007. No, the censors weren't asleep at the wheel— "to punk" now (also) means to mess around with, or to play tricks on. The word has a lot more "family-friendly" <cringe> play in it now that we're no longer (just) talking about the kind of tricks you "turn."

What's more, as tricks go, past-punking is one we're all equipped to perform. Imagination (everybody's got one, they're just not all put to good use) and history are inseparable. Recorded evidence may constitute the bricks of the historical mason's trade, but narrative is the mortar. Even the most vociferously "objective" historians make up a good deal of what they write. Herman Hesse said it before there was a "post-" to tack onto the modern: "We must not forget that the writing of history—however dryly it is done and however sincere the desire for objectivity—remains literature. History's third dimension is always fiction."[2] Need more license to mess with the past? Think of it as a gateway drug for social change. After all, it's a beautifully short step from "might have been" to "could still happen."

There's an incredible power in this idea. *History is a province of the human imagination.* In order to operate in the world, we need to imagine it, tell ourselves stories— give ourselves and the things and people around us roles in the tale of what comes next. It's no accident that the part of the brain responsible for memory is the very same part that sparks up when we think about the future. The paths we've taken set the lines on our map. From this point of view, steampunk is a kind of free base cartography—open-source orienteering dialed up to eleven.

'Course there's that elephant in the room. Whether we like it or not, steampunk has gained a lot of momentum in popular culture. It's cursed with a catchy name and an aesthetic that really resonates with our time-jumbled information age culture. That kind of fuel is more than enough to propel any subculture into the spotlight. Sadly, any good bohemian knows that cool is at best a double-edged sword. Past experience tells us that it's as like to cut your heart out as to spread your message. It turned Kerouac and Ginsberg (who punked each other) into "Beatniks," Kesey and Leary (who didn't) into "Hippies." Nasty shit, this "cool." Avoid the purple pills; people is havin' bad trips.

But our culture is changing. Media is multi-directional again (for the first time since it became "mass") allowing us to talk back to the cool-makers. And as far as subcultures go, steampunk is a pretty unique beast, lacking a lot of the burnout factors that have prevented other groups from mounting a resistance: it's not a youth culture—in fact, it's surprisingly multigenerational; it's not exclusionary—in fact, sociologist Mark Cohan has even remarked

---

1   *De Profundis*, (1897)

2   *The Glass Bead Game*, 1943

on its "radical inclusivity;" and it's not a drug culture—okay, a little absinthe here and there... point is steampunk is hardly gonna burn out on speed. Add all this up and, in the bigger picture, what does "cool" mean for those of us who retain some real faith in what steampunk is and can be? It ends up weirdly simple—it just ups the stakes.

I've spent the past year and a half talking steampunk with a huge variety of folks, and I gotta tell ya: it's got a mixed reputation. There are more than a few sharp minds out there that have serious reservations about what they see in steampunk. What's more, they've got good reasons—beyond just distaste for "hip" or "cool." Fact is, steampunk's romance with the past can be dangerous. Doubt me? Okay, I know it's fish-in-a-barrel, but pull up Justin Bieber's 2011 Christmas video (Santa Claus is Coming to Town"). Pause at 2:00. Now open another tab and pull up Wikipedia's entry on child labor (a similar effect can be achieved by watching said video then reading Upton Sinclair's *The Jungle*, it'll just take a lot longer and put you off sausages for a couple years).

Make no mistake about it, steampunking is a political act. Child laborers in turn of the century mills did not hip-hop dance on the job. Why not? No, not because "hip-hop wasn't invented yet"—this is my damn story—but because they'd run serious risk of losing a limb in the machinery. Then they'd be fired because the job required two legs. Where's the danger for us today? It's just too darn easy to paint up a skunk like a rose. Herein lies the rub (with a nod to Spidey and Uncle Stan): with great power comes great responsibility. When you play with the past, the balls you're tossing around aren't Nerf-ed. You're playing golf, not whiffle-ball. If you don't think about what you're doing, you're likely to crack someone's head open when you swing that driver.

I got the chance to chat with China Miéville at the Key West Literary Seminar a couple months ago, doing research for *Vintage Tomorrows* (my book and film project [the book is out now, film is still in the works]). If you know anything about him and his work, you can imagine how an hour across the table from that deeply insightful and razor savvy thinker can really bring a question like this into sharp focus. To Miéville, who doesn't think of his work as steampunk, but doesn't mind at all if others (like me) do, "steampunk is an aesthetic moment than can be diagnosed culturally like any other aesthetic moment and understood in terms of the political economy out of which it emerged." Take a look at our own historical moment and it's pretty easy to see why he finds no coincidence in "the precise temporal overlap between the rise of steampunk in the last... maybe fifteen years, and the rehabilitation of empire in Britain and the U.S."[3]

Gulf War, anybody? We have to own up to the fact that Victoriana is one of steampunk's golf balls. You just can't dig into the nineteenth century without butting up against empire in one form or another. Imperialism is some seriously dangerous shit. Not just "back then" but right the hell now. Handle with care. It ain't enough to say "we want to keep the good and toss the bad." History doesn't work that way. You can't strain the East India Company out of your cup of tea. I'm not tellin' ya not to play golf (well, not in this particular metaphorical sense at least—I retain all my doubts about the wisdom of shearing acres of land and pissing away metric tons of water to facilitate clubhouse networking for MBAs). What I am saying is: know what game you're playing. Remember to yell "fore!" and watch your swing.

If you don't want your steampunk creations to aid in the process of legitimating empire, you have to face it head-on. Fun and romance are all well and good, but don't go forgetting that nineteenth-century London high society (same goes for New York and Washington, folks—don't for a second think the Yanks are off the hook just 'cause the guys in blue killed more guys in grey) was built on plunder. Make no mistake about it, colonization is a messy, brutalizing process that exploits and/or degrades everybody involved. Gonna play with an imperial setting? At the very least pick up a copy of Albert Memmi's *The Colonizer and the Colonized*. It's a quick read, has been available in a good English translation for almost fifty years, and will do you and your audience a world of good.

Good steampunk isn't actually all that hard to do, but it does take awareness. Heck, the literary cannon is near bookended with the stuff. Michael Moorcock's 1970s *Nomad of Time* series grapples with exploitation, power, and violence throughout its brilliant meanderings. And just this year, Mark Hodder's *Expedition to the Mountains of the Moon* gives us another brilliant example of doing it right—full frontal glorious classic steampunkery, but with a deeply respectful, well-researched, multi-dimensional African setting that treats both colonizers and colonized in rich human detail. See? Totally do-able. You just gotta do a little homework.

So, listen to Oscar Wilde and *punk that past* (this is my story, remember? I jam words into the great epigrammatist's mouth with impunity). Wilde also said that "the one duty we owe history is to rewrite it"[4] (this time his words, my mouth; reciprocation being basic fair play). Messing about with your imagination is your god/dess-given right. But keep it safe, okay? Take the time to learn your clubs and balls before you start whacking away on the green. Remember that like "the personal," punking is political. Punk responsibly. Think before you punk. The past's got some sharp edges—you could put an eye out with that thing.

> Need more license to mess with the past? Think of it as a gateway drug for social change. After all, it's a beautifully short step from "might have been" to "could still happen."

---

3   Conversation with the author, 6 January 2012.

4   *The Critic as Artist*, 1891

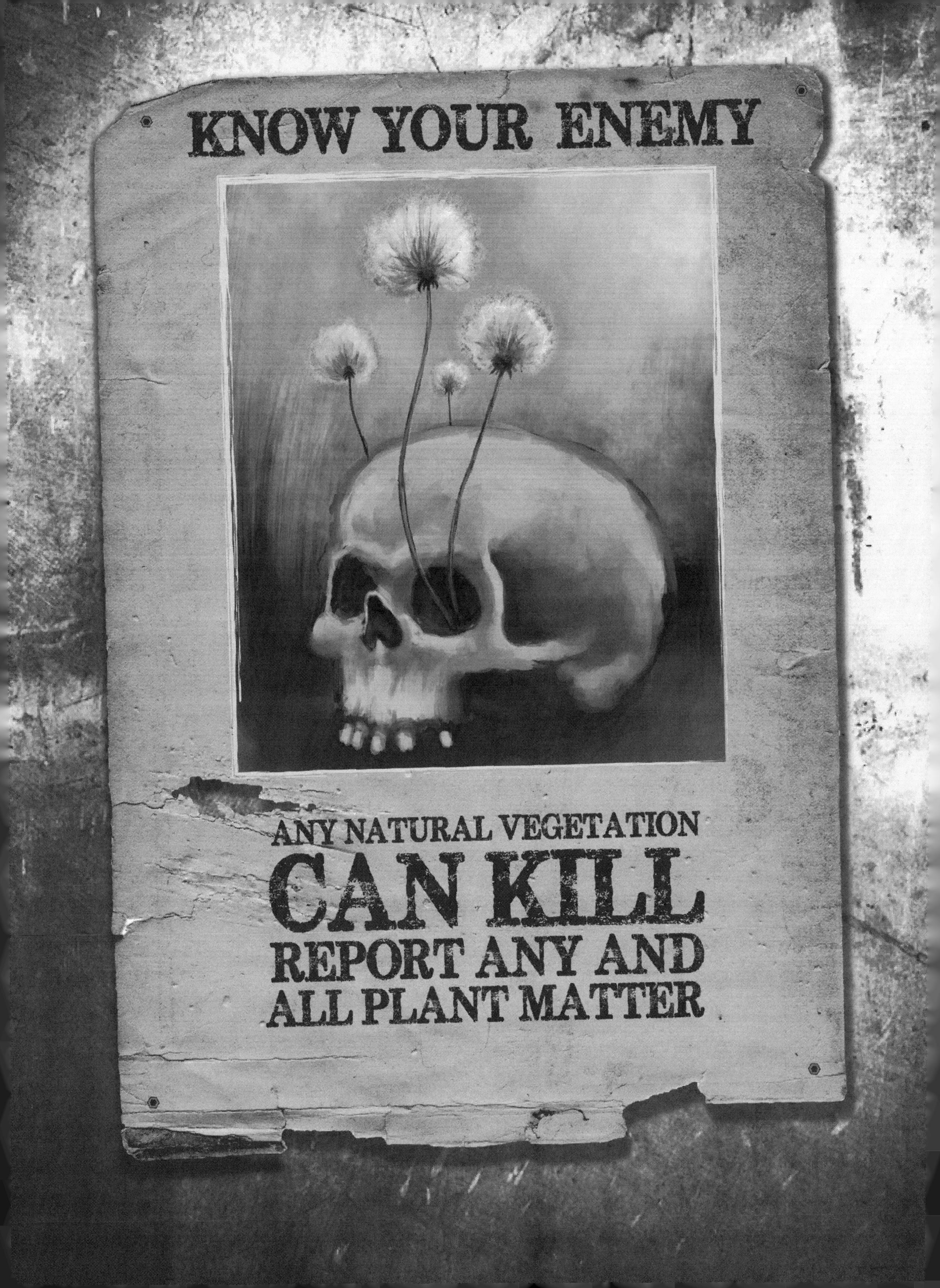
KNOW YOUR ENEMY
ANY NATURAL VEGETATION
CAN KILL
REPORT ANY AND
ALL PLANT MATTER

The Iron Garden

Erin Searles
Illustrated by Sergei Tuterov

The Gardener pushed closed the first of the heavy brass switches that brought the Iron Garden to life. Metal pinged and rattled as steam raced through the pipes of the root system.

He closed the second switch. The squeak of cables echoed through the underground chamber. Above ground, a meadow of elaborately hinged brass stems waved in time with the cables that snaked back and forth beneath them, their motion a mockery of the breeze. Carved jasper poppies and lapis lazuli cornflowers studded the golden field with colour.

The third switch started the belt that conveyed a queue of delicate metal birds to the top of the large iron tree at the centre of the garden.

At the fourth, the breeze bellows wheezed into life and conjured currents of air to stir the stiffened taffeta leaves.

The fifth switch opened a valve that sent gas speeding to the chandelier where the lonely pilot light ignited a parody of daylight. No sun filtered through the dome of blue glass and into the garden. Not anymore. The spore-smog hung too thickly.

The Gardener stumped up the stairs from the subterranean engineering room and into the Iron Garden. He turned his eyes towards the great iron oak—each leaf a chip of malachite hand-wired in place, the bark grooved with folds of metal, extrusions and handholds formed to give easy clamber to the young master's treehouse with its spyglasses and velvet rope balustrades. The Gardener held his breath.

The first of the brass-geared birds reached the very top of the tree. With a loud click, its retaining catch was unsprung and it took flight, spending the clockwork life the Gardener had wound into it beside the embers of yesterday evening's fire. The last step in a ritual of tending, where he would polish any dings and scrapes out of the burnished carapaces, stitch any rips in the oiled silk spread between the phalanges of the wings, and lubricate each gear. The mineral oil worked into every crease and wrinkle of his hand until his skin gleamed like paraffin wax.

The Gardener let go his breath in a long stream as he watched that first bird spiral above the Iron Garden. It was his routine to begin his maintenance rounds at the point where the first bird came to rest: his only shred of whimsy.

Elaine, his wife, had loved the clockwork birds. He had hoped, when he took the position as caretaker of the Viceroy's Iron Garden after the War, that bringing her here would make her happy. That it would bring her solace to be surrounded by intricate replicas of the plants she had loved so dearly—before they turned against mankind. It hadn't.

The bird's spring unwound its last twist and the cerulean silk wings stilled. The bird dipped into a glide. It sank out of view in a small copse of faux-birches. The Gardener gripped the worn-smooth head of his cane and made his way towards the copse.

He wove between the metal trees, searching for the bird. He peered around iron trunks covered in peeling silver foil and stared up through leaves made of wafer thin copper, mellowed to green with verdigris. He hoped the bird wasn't lodged in the high branches. The Gardener wasn't as young as he used to be. If he couldn't retrieve the bird, its cost would come out of his meagre pay.

With relief, he spied a flash of vibrant blue nestled at the base of one of the trees, close to the outer wall of the Garden. He hobbled over and stooped to lift the mechanical creature. The Gardener froze.

Yellow. Brightest yellow.

A garish colour not seen in the Iron Garden—where all hues were deep or rich or patinated. A flower sprang from a crevice between the base of a ferrous trunk and the deep-piled green chenille that served for grass. Not an enamel flower. Not a fabric flower. Not a copper flower.

A real flower.

Poison.

The Gardener's hand flew to the detector at his belt: a stripe of paper saturated in chemicals that would turn maroon at the presence of the poisons that spewed from infected flora. The paper was an expanse of unblemished cream, but the Gardener could not stop himself from trembling.

He had seen men poisoned by infected plants—those bastard children of the spliced strains, weaponised by the Northern Empire with common weeds and flowers. The deaths were not pretty. The plants' leaves soaked in oxygen and exhaled poison. Their seed heads ejected a smog of spores into the air that clogged each breath and burned the lungs.

A leak in the wall of one of the compound's workhouses had let in enough of the smog to kill three dozen. The Viceroy had ordered that section sealed off with men, women, and children still inside.

The Gardener leant heavily on his cane as he lowered himself onto his knees beside the brash, little flower. Its blunt, dimple-ended petals and vulgarly serrated leaves were at odds with the flowing lines and elegant aesthetic of the Iron Garden. His shaking hand detached the detector from his belt and held the paper a mere half inch from the leaves. No change. Not infected.

The Gardener let out the breath he had been clutching tight within his chest.

The yellow of the petals glowed. The flower seemed to soak in the artificial light of the Iron Garden and radiate it back with the gleam of reality. It bobbed in the air currents, its movements supple and artless, springing proudly upright whenever the breeze paused for breath.

Harmless. The little weed was harmless. Still, he had to report it. There would be an investigation of how the plant had gotten in and how it had survived in this environment. The plant itself would be destroyed with fire. He cast one last glance at it, then he snatched up the blue-winged clockwork bird, clutched it to his chest and bustled out of the birch copse.

He had to report this. The Gardener chanced a look back. The flower's sunny head bobbed in the artificial breeze, like a merry wave goodbye.

In his underground quarters the Gardener huddled under the light of the single oil lamp, his meal of thin stew only half eaten. His hands—clean as he could make them, but still grime-creased—turned the thick pages of the book. Elaine's book. The delicate pen and ink drawings of flowers so realistically rendered that merely looking at them made the Gardener uncomfortable. These fragile organisms had been an object of fear for all of his adult life. And before that… Before that, he remembered plants being neutral—just things that were there. He'd never really cared about them. Botany was Elaine's field; mechanisms were what fascinated him.

A memory rose up to submerge him and the Gardener closed his eyes. A lifetime away, he and Elaine sat on a tartan blanket bathed in sunlight mottled by leaves. Elaine wore a crown she had pieced together from daisies. In a high, sweet voice, she sang to him the names of the trees around them.

Quercus Robur. Fagus Sylvatica. Fraxinus Excelsior.

Stiff and upright, trying not to crease his newly issued military uniform, he nevertheless smiled to see her so happy. The grass was crisp and cool under his palms, the smell of it bright and alive.

The tiny clockwork man he had fashioned from scraps of metal toddled across the blanket, a ring with a peridot stone tied around its neck.

He had put the ring on her finger. She had tucked a yellow flower behind his ear. Its petals had tickled his skin. Their world was full of promise.

This was before the War, before the Splice. Before plants had been turned against man.

He reached the page he was looking for. Dandelion. Lion's Teeth. *Taraxacum*. The yellow wash that filled the head of the flower burnt from the page and into his eyes. It was almost as bright on the page as it was in the garden. The saw-toothed edges of the leaves stood out sharp as knives against the paper. As the Gardener pulled his fingertip down the page he expected to be cut. He wondered, if he touched the leaves of the real plant, would they feel soft or sharp?

The Gardener hurried across the garden. He passed through the orchard where the Viceroy's daughter and her friends sat on moss-coloured velvet tuffets. They used to put on tea parties for their dolls. Now they drank the synthetic tea themselves from real china cups and plucked hard candies shaped like apples and pears from where they were tied to the boughs of the wrought iron trees. The Gardener had spent the first part of the morning stringing the confections up with red ribbons; he was behind on his rounds and had a misery in his back.

He ducked under a branch into the copse of birches. The foil leaves didn't rustle and chatter today. The breeze bellows were off and the furnace stoked up to create the requested summer day for the picnickers. The trunks baffled the sound of the girls' trilling voices. A false quiet settled around the Gardener. As he knelt beside the dandelion his knees cracked like a jolting gear. His heart jumped up into his throat and came down racing.

His head darted around in furtive glances. He was unobserved. The Gardener reached out a gnarled hand and stroked the flower head. It was so soft he could barely feel it at all. He ran a finger down the centre of a leaf. It was smooth and yielded under his touch.

The Gardener peeled back the edge of the chenille floor covering from the base of the tree trunk. It revealed the crevice in the stone where the dandelion had taken root. The poor, thin dirt was dry—almost dust. He reached into his pocket for the small flask of water and uncapped it. He let a slow dribble of water out at the corner of the crevice, taking intense care not to wash away the dirt from around the base of the plant. He let the chenille roll back to cover the stone and made sure that all of the little plant's leaves were free.

He let himself feel a stir of pride at what the dandelion had brought forth from such weak feed. The hardy little plant did not really need his ministrations, unlike the myriad mechanical blooms of the Iron Garden. They required such pampering and tending; the dandelion stood tall with its proud golden crown and said to the world that it could take care of itself, thank you kindly. It found an opening, stuck its root in and made its own way in the world.

With a grunt of approval, the Gardener pushed hard on his cane to lever himself to his feet. He walked out of the copse and onto the path, where he almost crashed into the Viceroy's son as the boy tore along the pathway. The Gardener yelped, then remembered himself. He stooped his head in a bow and brought a knuckle to his forehead. The boy stared up at the Gardener with his wide, blue eyes. The Gardener stared down. Both waited for the other to move.

It was the Gardener who broke eye contact first. He grunted something—he wasn't quite sure what—and walked on. After a hundred yards he chanced a look back. The boy was still standing by the birches. Watching.

Behind the assembled guests, the Gardener shuffled his feet. They had been invited to see the unveiling of a new flower; he was there in case of mechanical failure. The Viceroy, resplendent in his formal attire, and his family stood apart from the rest. The ladies had clad themselves in rose-coloured gowns in honour of the day. The Viceroy's son wore full-length trousers and was fidgeting almost as much as the Gardener.

Mr. Romilly, the architect of the Iron Garden's newest wonders, concluded his speech with a bow to the Viceroy and to the audience, and a curt nod towards his assistants. They scurried to wheel away the screens from around the new bed to reveal three standard rose bushes. The stems and leaves of the bushes were reproduced in burnished bronze. At the edges of the tightly closed buds, a dusky pink enamel peeked out.

The crowd applauded: five pert taps of the hands each. Subdued murmurs ran through the audience, becoming excited chatter as one lady noticed the movement of the buds. The roses were opening. Romilly described the elaborate gearing that regulated the speed and synchronisation of the unfurling petals. So many flowers on each bush. The Gardener knew he would be fixing the accursed things every other day.

The first of the roses was now almost fully open. Mr Romilly's self-satisfied tones floated on to the pièce de résistance. The last petal fell open; a spray of fragrance diffused into the air. The crowd cooed and turned their faces up to take in the scent. The Gardener winced at it. The smell cloyed at the back of his throat. The artificial base left a metallic sting in his nostrils. Refilling the reservoir with perfume would be added to his list of daily tasks.

The Gardener ran his eye along the edges of the petals. Every line was of elegant design. The colours of the enamel selected

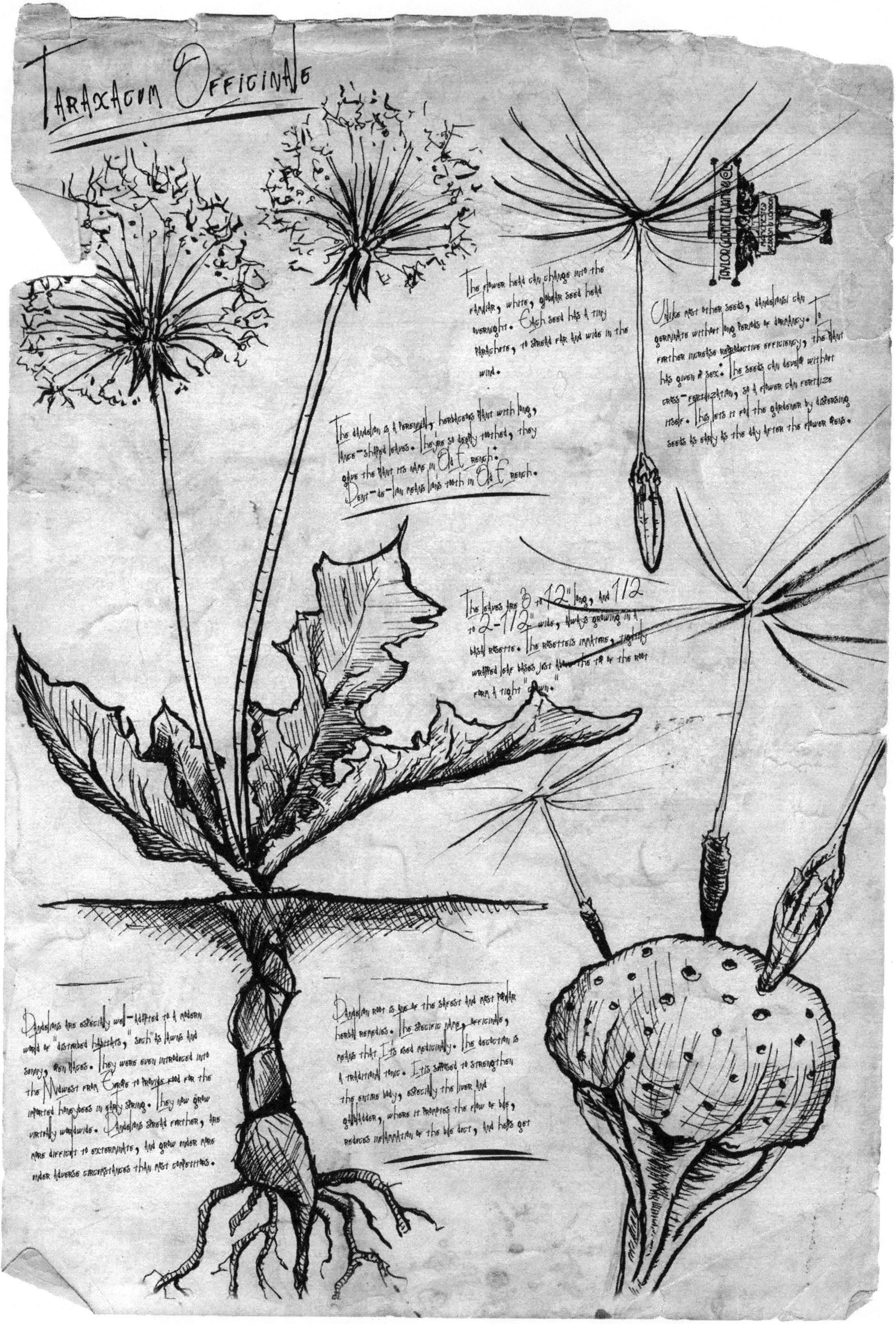
Taraxacum Officinale
The flower head can change into the
familiar, white, globular seed head
overnight. Each seed has a tiny
parachute, to spread far and wide in the
wind.
Unlike most other seeds, dandelions can
germinate without long periods of dormancy. To
farther increase reproductive efficiency, the plant
has given up sex. The seeds can develop without
cross-fertilization, so a flower can fertilize
itself. This lets it foil the gardener by dispersing
seeds as early as the day after the flower opens.
The dandelion is a perennial, herbaceous plant with long,
lance-shaped leaves. They're so deeply toothed, they
gave the plant its name in Old French.
Dent-de-lion means lion's tooth in Old French.
The leaves are 3 to 12" long, and 1/2
to 2-1/2" wide, always growing in a
basal rosette. The rosette is immature, tightly
wrapped leaf bases just above the top of the root
form a tight "crown."
Dandelions are especially well-adapted to a modern
world of "disturbed habitats," such as lawns and
sunny, open places. They were even introduced into
the Midwest from Europe to provide food for the
imported honeybees in early spring. They now grow
virtually worldwide. Dandelions spread farther, are
more difficult to exterminate, and grow under more
under adverse circumstances than most competitors.
Dandelion root is one of the safest and most popular
herbal remedies. The specific name, officinale,
means that it's used medicinally. The decoction is
a traditional tonic. It is supposed to strengthen
the entire body, especially the liver and
gallbladder, where it promotes the flow of bile,
reduces inflammation of the bile duct, and helps get

with discernment to complement perfectly the metal of the leaves. He curled his lip. The image of his dandelion pushed to the front of his mind. He smiled at its brash, unaffected confidence. He would visit it after this farrago was over. He was worried about the little plant. The flower head was starting to curl in on itself, as if in a sulk.

He turned with a snort away from the admiring crowd. He had duties to attend to.

The Gardener navigated around the silvery trunks of the birches. Their leaves clattered and clinked in the stiff breeze from the bellows. He hadn't visited his dandelion for days. As predicted, the maintenance and repair of the roses had filled his time. In those moments he did have free, he was often thwarted by the Viceroy's son and his playmates. The boys seemed to have decided that the best place to play marbles was on the wide stone pathway skirting the copse.

No flash of yellow greeted the Gardener as he approached the outer wall of the Iron Garden. The flower had been curled almost completely closed when last he saw it, but surely it hadn't died. He hurried his steps and reached the little weed. Not dead. Its sunburst head had turned into a bright, full moon. The Gardener couldn't help tensing in alarm as his thoughts turned to poison spores, but the detector at his belt remained clear. He reminded himself of the pictures in Elaine's book. These were the seeds. Not dying, but living. His dandelion was making more life.

As he left the copse, a glint of light flashed from the central tree. The Gardener blinked and moved on. He reached the formal garden before he realised that it was the reflection from the spyglass in the young master's tree house.

A scream echoed from the birch copse. The Gardener turned and rushed across the marble terrace, back towards the copse. He saw the Viceroy running from the gazebo, his wife hurrying after. Their son's insistent treble gibbered the words "spores" and "poison" and they doubled their pace.

The Gardener reached the dandelion and found the Viceroy shielding his wife and son from the small plant. The boy clung to his mother, spluttering out breathless sobs. One wet, blue eye peered out from over her shoulder and narrowed in malice at the Gardener.

The Viceroy barked an order about oil and matches. The Gardener opened his mouth to protest. He held out his unblemished poison detector. The Viceroy's face turned maroon; his shouts increased in volume and panic. The Gardener closed his mouth with a snap. He fled.

The oil and matches lay on the bench in his underground workshop. The Gardener picked them up. He was surprised to feel a tear forming at the corner of his eyes. Through the blur, he saw the controls to the breeze bellows.

With the bellows on full, the copper-foil leaves rattled in agitation. Only the Viceroy remained in the copse, his sleeve over his mouth, his face drained pale. He clutched his poison detector in front of him like a talisman.

"Hurry up. Hurry up, man. Destroy it."

His tone sped the Gardener's feet. As the Gardener hurried forwards, he let his foot catch on a root. He fell, and his abrupt landing stripped the dandelion of its seeds. They exploded from the plant's head and hung for a moment in the space before the Gardener's face.

He watched as the seeds were picked up by the currents of air coming from the bellows. They danced.

The Viceroy yelled in anger. He tried to swat the frolicking wisps from the air with his gloved hands. The floating seeds merrily evaded him and slipped away into the Iron Garden beyond.

The Gardener hoped that at least some of them would find an undisturbed crevice to nestle in. Elaine would have liked that.

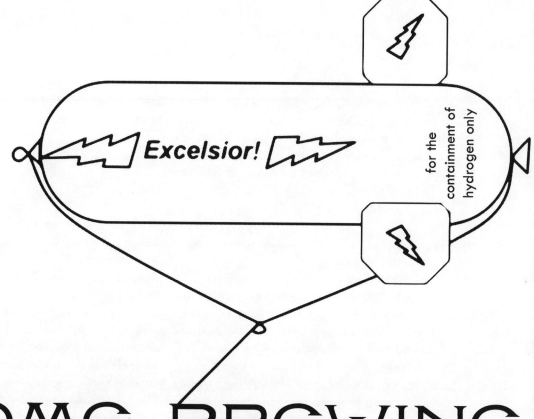

Home Brewing Miniature Vegan Airships

By Professor Offlogic

Important Safety Precautions:

- **Read the entire article and all safety notes twice.** Think about how good you are at following directions. Remember that time when you mistook teaspoons for tablespoons and the muffins you made were like rocks? Read all of this again. Do you think you'll probably hurt yourself? Think about it long and hard before going forward.
- **Do not make hydrogen indoors.** Hydrogen forms explosive mixtures with air over a very wide range of concentrations and these mixtures can ignite with the slightest spark. Even a small quantity of hydrogen going **BOOM** can produce a shockwave inside a room sufficient to shatter windows, personal injury all aside. Outside, any leaked hydrogen will quickly disperse. Just don't do it inside.
- **Static electricity is very sneaky and dangerous!** Don't handle hydrogen on cold, dry days optimized for static electricity. Avoid wearing synthetic fibers, silk, or wool when handling hydrogen. Bare feet on grass or bare earth would be a prudent precaution (although working nude might draw unwanted attention).
- **There should be no possibility of sparks or flames in the immediate area.** No electric lights or light switches, motors, doorbells, Tesla coils, candles, oil or carbide lamps, torches, pilot lights, or flames anywhere near the hydrogen. Needless to say, smoking around hydrogen is more hazardous to your health than usual.
- **This project involves passing electrical current through water, and that spells (manageable levels of) DANGER.** Never put any part of yourself in the water with the power supply on or the electrodes connected.
- **Proper eye and hand protection (goggles and gloves) would be wise.** Get adult supervision. Eat your veggies and get some exercise, you look so pale! Wait one hour before swimming and remember to floss regularly.

The history of the lighter than air flight dates back to Francesco Lana de Terzi, who first posited the concept of the "vacuum airship" around 1670, though, of course his design was never (officially) realized, he is cited by many as the "Father of Aeronautics". Still, it wasn't until 1784, when Jean-Pierre Blanchard began experimenting with manually operated oars, flapping wings and propellers on the gondolas of hydrogen balloons to achieve powered flight, that the (duh-duh-duhhhh) *Age of Airships* began!

For most of their history airships had been status symbols of colonial powers; who else but a great empire like Germany, Britain, France, Italy (and that upstart, the United States) could afford to build such extravagant aerial monuments to their power, prestige and technological know-how?

In these days of modern times the home experimenter is able to easily construct an apparatus for producing hydrogen lift gas using only common household items and simple hardware. With the methods and apparatus described below you will be able to launch your own miniature airship fleet.

Hey, air pirates have got to start somewhere, *n'est pas?*

The Lightest Element

Hydrogen, composed of only one proton and one electron (disregarding isotopes like deuterium, tritium), is the most abundant element in the observable universe and the lightest element known. With only 1/14th the density of air, each liter provides about 1.2g of lift.

A small molecule, hydrogen is much more nimble than its massive cousins like oxygen and nitrogen, able to diffuse through materials and containers that are sufficient to imprison these heavier gasses. Once free of its confines, hydrogen will do its level best to rise to the upper regions of the atmosphere to happily join its interstellar hydrogen brethren, born away on the breezes of the solar winds. Confined, hydrogen is sullen and resentful, spoiling for the chance to rejoin oxygen in a union that starts as fire yet produces purest water. The consummation of hydrogen's desire can be very loud indeed! All it needs to achieve this end is a single spark....

It is hydrogen's guile, cunning and fiery nature that make generating hydrogen an activity best pursued outside the household, where any leakage can freely disperse. This is the main safety precaution your should follow when brewing hydrogen at home.

Generating Hydrogen

During the American Civil War observation balloons were fielded with gas generator wagons, in which strong acids or bases would be combined with metals to generate hydrogen. In the last decade, scientists at Purdue University have discovered that aluminum, when combined with the liquid metal gallium, will rip the oxygen from water and release its hydrogen spontaneously. Who doesn't love fresh hydrogen?

Since the bulk of the hydrogen supply on planet Earth is tied up in water, in which two hydrogen atoms and one oxygen atom are bound, vast quantities of hydrogen are available in the rivers, lakes, and oceans. Since we live in an Electrical Age, it is only fitting that the power of electrical current be used to free these gasses and bend the hydrogen to our own uses.

Selecting The Envelope

Through the first quarter of the 20th century most airships relied on *gold-beater's skin* (a specially prepared form of cow's intestines) to contain their lift gas. Later, cotton cloth sandwiched with gelatin was used by some. Our miniature versions will be of a more vegan type, relying on latex or modern polymers for gas containment, however temporary.

You may elect to use toy balloons or thin plastic bags (e.g. dry cleaning bag) for your miniature airship envelope, but bear in mind that the bigger the envelope the longer it will take to generate enough hydrogen to fill it. You want the smallest, lightest envelope you can get and still have fun with.

Ordinary latex condoms will make a usable envelope, and offer several advantages: condoms are produced under tighter quality controls than toy balloons, require less pressure to inflate, are roughly airship-shaped, come in a variety of colors and textures, and their water-based lubricant helps reduce the dangers of static buildup in the envelope itself. While I have not actually tested "latex free" urethane condoms, they are reputed to be not as amenable to inflation due to a lack of flexibility.

The Trojan "Enz" condom, for example, is 2 inches wide and 7.5 inches long "empty" and weighs a scant 2.5g. Filled with 2 liters (just over one half gallon) of hydrogen they can achieve neutral buoyancy, but can be further inflated with a full 20 liters (slightly over 5 gallons) to provide over three quarters of an ounce of lift.

A Note on the "Threat" of "Deadly Chlorine Gas!!!"

In almost every discussion of the electrolysis of water you will encounter at least one earnest soul warning all to beware the Lethal Clouds Of Chlorine Gas that will inevitably be produced when one uses ordinary salt as the electrolyte.

Let me be quite clear: those making these claims may be forthright but suck at risk assessment, or they may be simple slackards that never shifted a cheek of their indolent arses to actually do anything, yet wish to sound authoritative.

Ignore the cackling of poltroons, these nabobs of do-nothing! Salt is the electrolyte used here because it is cheap and safe on bare hands (unlike the sulfuric acid or caustic hydroxides often inveigled by lumpish whackaloons). Absolutely minuscule amounts of chlorine will be produced during the electrolysis of water as described herein. You'll get more *"Deadly Chlorine!"* exposure just splashing around in a hot tub.

On balance, you are far more likely to electrocute yourself with the battery charger than recreate the trenches of 1915 Ypres in your backyard. Idle scare-monkeys can *snort my taint*.

The Electrolytic Hydrogen Generator

Materials:
- A car battery charger or other low voltage power supply. 5 amps or better would be nice, but it *must* be a direct current supply (car battery, solar panel, etc).
- Plastic tub or bucket (not metal, PLASTIC) and several plastic milk jugs with caps
- Stiff insulated copper wires (solid is best, like Romex house wiring)
- Salt
- Water
- Misc. aquarium air tubing, bulb-type syphon pump, glue/caulk and etc.

Setting up the generator:
1. Put several inches of water in the tub or bucket. Add 4 tablespoons of salt per gallon of water.
2. Completely fill the bottles or jugs with water. Add 4 tablespoons of salt.
3. Cover the mouth of the jug (either with the palm of your hand or their caps) and burp as many air bubbles out as possible. We want water, but no air, in the jug.
4. Invert the still-covered jug into the tub with the mouth under the water level in the tub/bucket.
5. Uncover the mouth of the jug. The jug should remain water-filled without drawing in air.
    a. If it glugs in air from the mouth, increase the water level in the tub and repeat steps 2-4.
    b. If it draws air bubbles from anywhere else, discard the jug—it leaks!

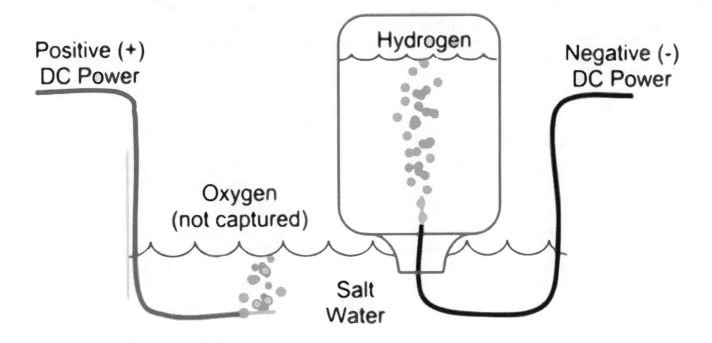

6. Secure the jug so it will remain in position with the mouth below the surface of the water (the solid insulated wire can help with this). If filling several jugs, attaching them to each other may enhance stability. Bear in mind that as hydrogen is generated the jugs may attempt to float. As long as they stay inverted with their mouths below the surface of the water, all is well.
7. Repeat steps 1-5 above for as many jugs as you intend to fill at one time.
8. Prepare positive (+) electrode wire:
    a. Select a color to use for positive (+) power supply connection (RED is shown). If only one color of wire is available, label this wire clearly.
    b. Cut enough of the (+) wire to comfortably reach from the DC power supply (this would be the (+) alligator clip of the car charger) into the salt water in the tub. Strip several inches of insulation at both ends of the wire.
    c. Use tape, clothespins or plastic clamps to secure the positive wire in position.
    d. Do not connect the power supply at this time.
9. Prepare negative (-) electrode wire:
    a. Select a color to use for negative (-) power supply connection (BLACK is shown). If only one color of wire is available, label this wire clearly.
    b. Cut enough of the (-) wire to comfortably reach from the DC power supply (this would be the (-) alligator clip of the car charger) up into the salt water in the inverted jug. Strip several inches of insulation at both ends of the wire. *(Pro Tip: If the bare end of the negative electrode wire is positioned inside the jug just an inch or two above the mouth, electrolysis will automatically stop when the water falls below that level).*
    c. Use tape, clothespins or plastic clamps to secure the negative wire in position.
    d. Do not connect the power supply at this time.
10. Do a final safety check:
    a. Are the jug(s)/wires going to stay put?
    b. Is your power supply positioned where it can't get wet should the tub overflow or otherwise have water spilled? Is it at least 6 feet away from the hydrogen generator?
    c. Are you still smoking? Put that cigarette out! Extinguish all flames!
11. With the DC power supply turned off, connect the positive and negative electrode wires to the supply. *Remember: while the hydrogen generator is running there is electricity flowing through the water, and that means* **KEEP OUT OF THE WATER WHEN THE POWER IS ON**. It's a good idea to keep one hand in your pocket when working with electricity anyway.
12. After drying your hands well, turn on the DC power supply. You should begin seeing bubbles coming up from the electrode wires in the salt water immediately.

a. Those bubbles at the negative wire are the hydrogen, collected by the jug.
   b. The bubbles at the positive electrode are oxygen. Let those just drift away, but make sure they don't bubble up into the hydrogen jug.
   c. If you don't get bubbles, *turn off the power* and then try adding more salt, decreasing the distance between the positive and negative electrodes (mindful to not let the oxygen bubble up into the jug) or exposing more bare wire at the ends of the electrodes, coiling it up is just fine. Recheck your electrical connections SAFELY (with power off).
13. Now is the time to wait and watch. As hydrogen is generated in the jug it will displace the water into the tub.
14. Once the jug has reached capacity or the target amount (or stopped generating automatically if you followed the tip in step 9), turn off the DC power supply THEN disconnect the electrode wires (in that order, no electrical sparks, remember?).

Electrolysis can take quite a while depending on your configuration. Larger electrodes placed closer together (being careful to keep the oxygen out of the hydrogen), higher concentrations of salt and higher current levels can speed things up, but it always seems to take forever. Don't leave the generator completely unattended, as raccoons might play with it.

Filling the Envelope
You've made the hydrogen, now you have to coax it into your chosen envelope.

If you have opted for a non-pressurized envelope (such as a dry cleaning bag, produce bag, etc) the filling process is rather easy, since hydrogen "pours" up. Simply righting a jug with the envelope held tightly around the mouth (to reduce air infiltration) will let the hydrogen rise into the envelope.

Filling a pressurized envelope (such as a toy balloon or condom) requires mild pressurization to overcome the resistance or the latex. Since hydrogen is a slippery little devil, pumping it directly is a tricky business. The simple pressurization scheme below uses water to accomplish the task.

1. Drill one ½" and one ¼" hole in one of the jug caps. Press-fit the outlet tube of a small bulb-type syphon pump (generally used to siphon gasoline) through the larger hole. Fit a length of ¼" tubing such as aquarium air tube or drip irrigation tubing through the smaller hole. This will be the fill tube for your envelope, and should be long enough to reach up to the bottom of an inverted jug. Secure and seal around both tubes with adhesive or caulk ("shoe goo" works fine), making sure the cap is still able to snap onto the jug.
2. Completely flatten the envelope, closed end-to-open end, to expel as much air as possible. If using a condom, fully unroll it before flattening. You may wish to inflate your envelope (using something other than hydrogen) and decorate it with markers before deflating and re-inflating with hydrogen.
3. Affix the open end of the envelop over the fill tube to achieve a tight seal. This may be facilitated by first forcing the outside end of the fill tube through a close fitting hole in a spare jug cap (particularly true of condom-base envelopes). Rubber bands may help.
4. Carefully work the cap and tubes under the water and snap it onto the mouth of a jug. The fill tube should be standing up into the highest part of the jug, with the collected hydrogen. *Important Note: the hydrogen will try to escape through the fill tube as soon as it is introduced into the jug, so the outside end must be closed with the envelope or pinched firmly shut during this step.*
5. While holding the envelope in place on the fill tube, work the pump to force salt water into the jug. Hydrogen will fill the envelope.
6. Repeat steps 3-4 for as many jugs of hydrogen as you deem fit to use.
7. Twist the open end of the envelope closed, remove from the outlet hose and tie off securely.
8. Attach a string and start flying! If using a condom, tying a string between the reservoir tip and the aft knot will allow you to find a tether point that causes your miniature airship to fly straight and level. Keep moving the tether toward the low end until you achieve proper trim.

Embellish your airship! Attach fins, gondolas, gun decks made of light card-stick! Hoist an LED and coin-type battery! At higher levels of inflation, you may be able to attach rubber-band engines to achieve truly powered flight! Stage epic air battles between the forces of Good and Evil (I'm putting a few shekels on Evil). The sky and your imagination are the limit!

If going for high altitude in free flight, remember that the lift gas will expand with increased altitude. You should leave your envelope only partially inflated lest the expansion cause "burst height" to be reached prematurely.

Just remember while operating your new fleet that static and flames remain hazardous! Best to keep your aerial dreadnoughts at the end of several feet of string than get an up close and personal demonstration of what happens when things go **horribly wrong**.

You have been *warned*. Now go have some mildly hazardous fun!✺

Bulb-type
fluid pump
To
Envelope
Hydrogen
Salt
Water

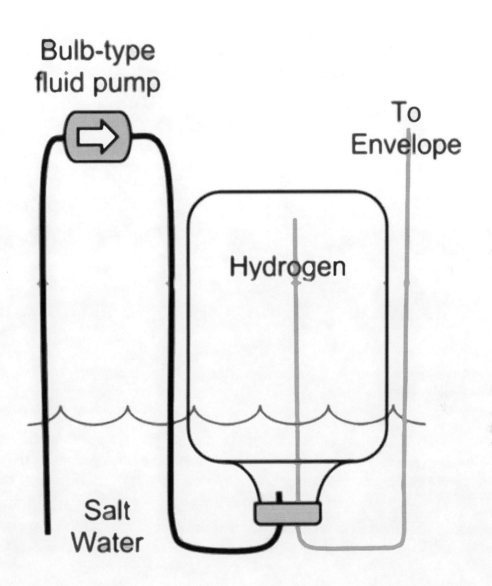

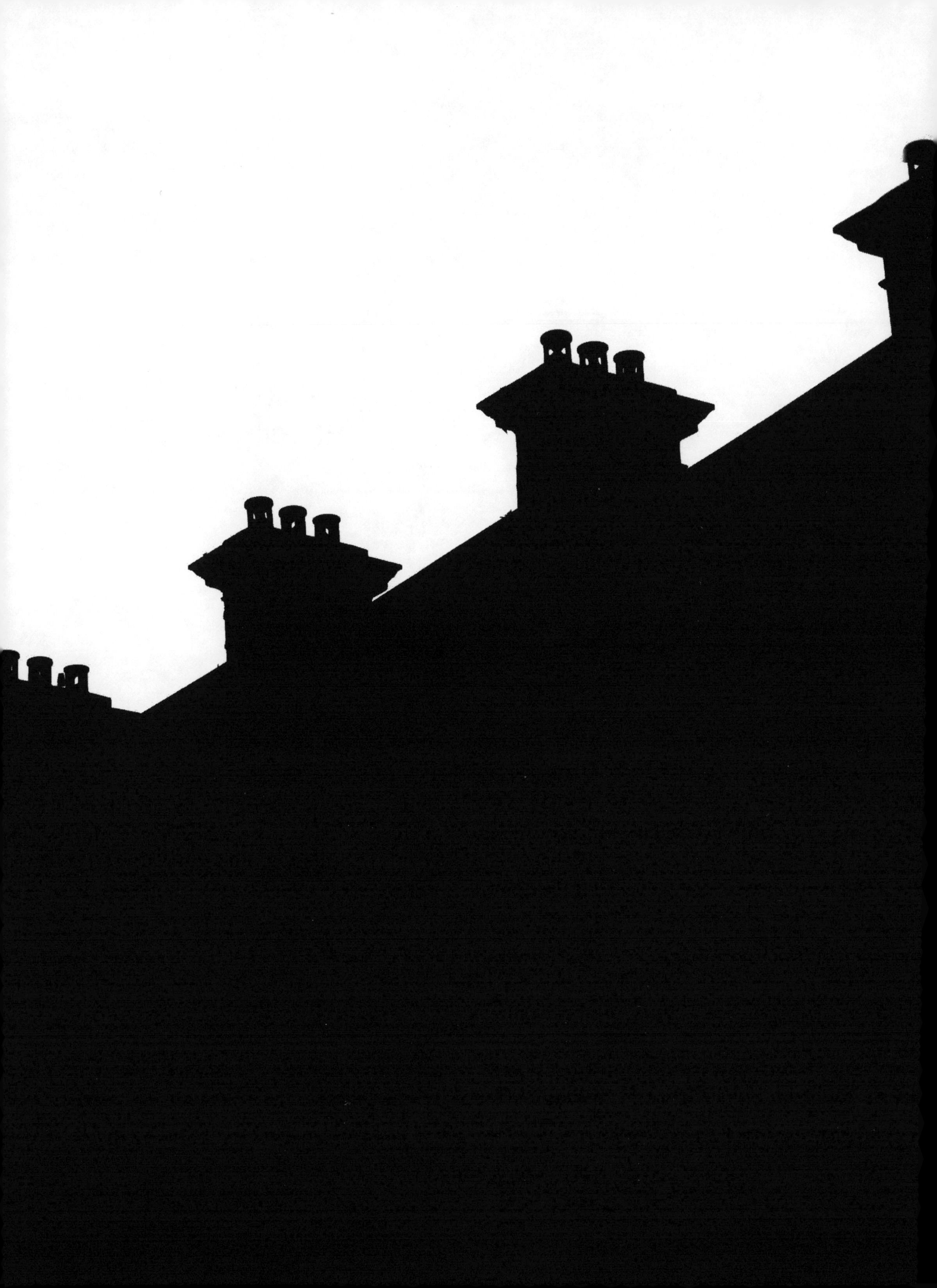

Hell's Dinner Plate

by The Catastrophone Orchestra

April 30th, 188?, New York City

After exiting the overflowing privy in the courtyard of their squatted tenement, Pip allowed himself to breathe. But as he was folding a three-day old *New York Tribune*, a shower of small glass bottles forced him to retreat back to the stinking privy. He waited until the sound of broken glass was replaced by shouts from his bandmates before emerging again. His goggles were tight over his face, the only thing keeping the stench from making his eyes water.

"A man can't even have a peaceful morning shit in this madhouse," Pip grumbled to himself.

Pip wasn't surprised to see a group of filthy Lower East Siders waiting uncomfortably in the main parlor/waiting room. He carefully stepped around his tools, gaskets, gears, and other ephemera that lay strewn across the threadbare carpet, then took off his goggles and hung them on a lever of his greatest invention—the Catastrophone. The steam-powered instrument was still mangling its B flats, and Pip was desperate to get on with his tinkering so the machine could be loaded for their show that night. But he couldn't concentrate with those three grimy mugs staring at him. He pulled open the ripped screen separating the kitchen/laboratory from the parlor and stepped through.

Neal List was complaining while Professor Calamity was poking around under the sink. The professor looked like a grasshopper—all long legs with his head buried underneath the sink.

"How can a self-respecting clinic be out of laudanum?" Calamity asked, standing up and shaking an empty bottle.

"First, we ain't self-respecting," Neal started, counting off on his fingers. "Second, we need more than handshakes to buy that blasted stuff of yours. Third, you practically guzzle it anyhow."

Calamity started to clap slowly.

"My dear Pip, look how Neal has mastered the confounding and subtle principles of arithmetic. But despite his considerable mathematical prowess, the methods of a community clinic seem just beyond his impressive cerebral reach. Our patients wait. Do you have any coins so Mr. List can hop to the apothecary and procure what we need?"

Pip checked his pockets, though he knew he was flat skinned like the rest of them. "Credit maybe?" he suggested.

Both Neal and the ex-alienist started to laugh.

"That chemist has already sent his shoulder pushers around twice this week to flatten the tab," Neal said.

"And we cannot very well send our patients into Hell's Dinner Plate without the necessary prophylactics," the professor added.

"Besides, we need some money to move Catastrophone for the show tonight." Neal said.

"This is truly confounding," Calamity said, rubbing his thin needle-marked arms. "I just can't think straight. Get my love."

Neal and Pip exchanged uncomfortable looks while the professor slid to the floor, holding his knees muttering to himself.

"Fine. I'll wake Mathilda," Neal finally said, steeling his nerve.

In a moment, Pip peeked through a rip in the screen to see if the patients, a chimney sweep and his child apprentices, had fled with all the hissing coming from the bedroom—waking Mathilda was not for the faint of heart. But the poor bastards were still there. The people that came to clinic came because they had nowhere else they could afford to go.

Screaming Mathilda, an ex-patient of the professor when he worked at Bellevue, entered. Calamity flew from the floor into her outstretched arms.

"My sweetest," Mathilda cooed pulling him close, "I need my sleep before the performance."

"I have no laudanum. Candlestick and his squirrels need some ointment. They are even willing to pay. How can I make the medicine?"

"You are a healer. You healed me. What do you have?"

"Everything but the laudanum," Calamity said, planting a kiss on her lips.

"There ain't no way to stop soot wart," Neal whispered to Pip while they waited for the two to finish kissing and groping each other. "Those kids are as good as gone, no matter what."

"Did you not tell me gin had miraculous healing powers?" Mathilda asked, framing the professor's thin face in her pale hands.

"Did I?"

Mathilda pulled up a flask from her garter. Calamity turned to the sink and mixed the ingredients with the gin, taking a swig.

Pip went to give Mr. Candlestick the patent medicine. He spied one of the man's apprentices, a little black boy, hiding Pip's goggles in the pocket of his long coat. The eleven year-old caught Pip's stare.

"Here's your medication, Jack," Pip said, turning to the master sweep. "Seventy-five it is."

Candlestick passed him coins and Pip pocketed the change, "You know nothing in this world is going to keep those kids safe. This might buy them a bit of time. But if you keep sending them down, they'll get sick or worse. You know that, right Jack?"

"I'm workin' on it."

Even with the soot and grime, Jakob Leuchter was convinced he could still smell the corpse on his jacket. He seldom wore it, but today he needed to get at least a nickel for Schimel's gerkins, and Lev was the customer to go to for that. So he checked all thirteen of his copper breast buttons to make sure the thread would hold up to the tugs and rubs of the day, to make sure he looked his best. The yipsels were sleeping behind Meyer's pigeon coop three stories up. He shouldered his brushes and stepped out into the street calling.

Jack knew a number of callings. He could call out the night-soilman song or the bone-man's or even that of the pure schlepper. He had collected shit, rendered animals for bones for the button-mongers, and sorted threads by color from collected rags. He was chumitzer, a scavenger. He was one of a legion of men, women, and children who removed the mountains of the old, the impure, the broken, the overlooked, and the discarded. Yes, the city had the Commission of Sanitation with their tight white tack hats and polished green wagons of broom men that represented the face of cleanliness, but they were nothing. They were just another Tamany work scheme to make people feel like the city could actually function as long as they voted for Boss Tweed.

Jack knew it was schmucks like himself who kept New York from drowning on its own filth. People like him did not get starched white uniforms or weekly payoffs from the Irish Tiger. People like Jack wore other sorts of uniforms. The sweeps and their yipsels wore the secondhand tails given to them, by tradition, from undertakers. These tails were used for viewings and were provided by funeral houses to families who could not afford to waste perfectly good clothes on a corpse. For fifteen cents they would rent the grievers a set of tails for the viewing, which would be removed afterwards. Jack didn't know where the tradition had come from or why it persisted, but undertakers continued to set aside old coats for the sweeps. Thirteen buttons, the number of death.

People liked to dart towards Jack as he sang his song. They would grab a button, flash a smile, and return to their errands believing good luck had chanced upon them. Some even dropped a penny in his pocket, and this is what he was hoping for. The superstitions surrounding the sweep confounded him. There were no such traditions around the mudlarks and blood-collectors or around any of the other jobs he'd had before this one.

The luck of the sweeps, he thought, *tell that to the yipsels who struggle for breath like old men by the time they are nine.*

"You wear a dead man's tails in my home?" Lev said, waving Jack to pull up a crate.

"But I bring you gerkins."

"Schimel's?"

"Of course. The best on Ludlow Street," Jack answered, pulling the pickles out of his pocket.

"He charges too much, you know. But they *are* good," the old man said.

"Should I set up the board?"

Jack didn't wait for an answer, shoving the swollen pickle in his mouth before fishing out the chess set.

"How are your skuir-rels?" Lev asked, wiping the juice from his grey beard.

"It's squirrels. How come you always use the English word for them? Everyone else calls them yipsels."

"They aren't Yid," Lev responded.

"Neither are schmucks."

"Unfortunately, most of the ones I know are. So I use… squirrel… same thing as yipsel. Why bother? You let them call you Jack Candlestick. Jakob Leuchter was your father's name. Candlestick you sound like one of those vaudevillian clowns. It is your life to live and I am an old man who is still waiting for his wisdom. Now, since you are wearing a corpse's suit, you can be black."

Jack didn't want to believe Lev was racist. In fact, he had nearly convinced himself that Lev couldn't be a bigot because he held nothing in common with the big mouth nativists that terrorized the Five Points. But Jack had been in the world long enough to know it wasn't just the nativists in their star-spangled vests that hated his apprentices because of their skin. He knew that prejudice came in a hundred different forms in this ignorant city. He'd seen the omnibuses rumble past a mother and her children waiting at a stop simply because they were black. But what Jack really couldn't stand was the ignorant talk that surrounded him daily. When he'd first gotten back to the city, he'd sometimes gotten filled with whiskey and fought those big mouthed, no guts men. To most of them, a Yid was no better than the blacks. But people didn't know Jack was Jewish because he was exceptionally tall, remarkably skinny, and ready for a fight.

Jack then noticed Lev's shaking while moving a pawn.

"You still take the medicine?"

"That doctor is a shyster. He raised it to a buck fifty. What nerve."

"I have a big job tonight. I can get it."

"Posh. Those damn things didn't do nothing but make me piss like a racehorse. Save your money. A little shaking is to be expected when you get to my age, God willing. Give it to your yipsels. How many you got now?"

"Just Chase and a new girl called Rosie. Malcolm left last month. He saved enough to rent a cart. He does alright schlepping for the boys at the market on Sundays."

"Who's going to save you, Jakob?"

"I do right. Got these fancy clothes don't I? I'm ready for the opera with this hat and tails."

Lev's laugh turned into a cough that caused his face to flush. By the time the old man was done coughing, tears were in his eyes.

"We'll play some other time," Lev said still trying to catch his breath.

"I'll come around tomorrow morning and give you a little. It's Jack o Green tomorrow. I got the whole day off. We can go to Edna's for a bite."

"No, save it. Give it away. No need to waste your money on me. I'll be here a bit longer to beat you in chess. Just don't lose that coat. I might need it someday soon. So what's this big job?"

"That Vaux building. The Metropolitan Museum they're calling it. They say they got mummies and all sorts of stuff inside. It's supposed to be real flash."

"What are you doing there? I read somewhere that they were using those new whatchamacallits at that building. All science and progress."

"Yeah, it's a bunch of bunk and even they know it. They got some egg-heads with some kind of machine that supposedly can clean chimneys. Same ballyhoo every year. Reformers hold a contest and some crank comes up with the new brass dingle to clean the hell's plate so kiddies don't got to do it no more. I'll believe that when cops ain't crooked. It ain't worked yet and it never will."

"It's a nice idea though. I guess that's why they keep trying."

"Nice ideas don't keep your chimney from catching fire. So they got us. We'll go up and it will be another night's work for us when those flashy toys fail. Just another dirty job. The only difference is they are paying a fiver for each chute."

"It must be the Taj Mahal." Lev said, his breathing less labored.

"It's something, that's for sure. They say the museum is gonna be free."

"They say all sorts of things. We'll see."

Jack picked up his pace when he saw Goony and a bunch of his yipsels jump across the spans to "his" roof. The sun was already setting and the top of the Lower East Side was all swirling orange and red like tutti frutti ice.

"What are you doing noising around here, Goony?" Jack demanded.

"Is dat anyways to bark at an ol' college?" Goony asked, striking a match on his shoe.

"It's colleague, you ignorant tunnel rat."

"Whooa. Look at him now kids," Goony said. "This gent is Jack Candlestick. Tall 'nuff to poke God if he wanted. Sure as Shirley. Me an' he were *colleagues* once upon a time ago. Logan's Creek wasn't it? Can't right remember, all those hick mule towns kinda blur together when Johnny Reb is shootin' atch ya."

"I asked you what you are doing here. On my roof." Jack looked over at his two wards—they were still laying on their soot sacks but they were wide awake.

Jack didn't listen to the crude whisperings of Goony's crew. They were just kids, like all the yipsels, but they were a nasty bunch. He kept his focus on Goony.

"Tired of bendin' my hump in the sewers lookin' for bits, waitin' to get blown to tenny bits from gas. Thought I toss my hand in the fire prevention racket. Got me this lot for two fifty. Got to bring 'em up right; God, country, and the rest. Look at these buggers—ever lay your eyes on a finer bunch?"

"How old is that one, with the scar?"

Goony shrugged and shoved his thumbs in his greasy suspenders. Jack knelt and asked the undernourished and pale boy himself.

"That kid isn't even five! There are laws. Do you even got a license Goony?"

"You see a copper around? Those lard bottoms couldn't get up here with a hot air balloon, Jacko. 'Sides, I tell everyone you vouch for me, since we're pals and all. Just send a runt down and up quick as can be. It's easy money. We could even be partners like 'fore. Goony and Candlestick. 'Cept none of those," Goony said, nodding in the direction of Chase and Rosie. "I know they must be a dime a piece but I won't work with their kind. I gave them animals six months and you gave them your knee."

"We deserted in four months. Don't you worry about my knee. You don't know a thing about chimney sweeping."

Goony spit over the side of the building without looking at who might be below.

"Ain't no one knows lick spit. Pop down, get the soot, and back up. Easy breezy."

Jack knew that, in a way, Goony was right. The so-called master sweeps, those that carried licenses in their corpse jackets, not one had ever been down a single chute. The life of a sweep was short and few if any of the kids who scampered inside the chimneys ever reached an age to have their own crew. Jack was no different. He had never put more than his head in the opening of a chimney.

"Got a big job tonight," Jack said, "so I got to cut this reunion short. I do want to give you a piece of advice though."

"Goes ahead. You're always the one with the ideas… not for you I would have laid in some hick ditch with a belly of lead."

"Go back to the sewers. It's a swindled game up here. No mark up, with keepin' the yipsels safe and fed."

"Ahhh, just like a Jew, always too shrewd," Goony said, pointing his cigar at Jack, "'Fraid of a lil competition, you greedy bastard. Hey, but you can tell an ol' pal where you get dose clothes. Smart they are."

There was a pulley hanging like a spider off the museum's roof, and Jack had lugged the cane ladders eighty blocks for no reason. Still, he couldn't wait to get up on the roof.

Jack had been "stomping the shingles" for three years and one of the few things he liked about the job was the quiet—solitude was hard to find anywhere on street level in a city like New York. The only other thing he enjoyed was when one of his ward's hung up his brushes for the last time and Jack gave them a key for their box down at the Manhattan Trust.

He kept a box for each of his wards. He only had a few, especially compared to other squads, but what he'd told Goony had been true—it cost more than half of his fees to buy food, baths, medicines, and gear for his yipsels. Even with the extra cash he picked up selling the soot to farmers in Long Island, he barely had enough to keep a roof over their heads. But when he had some extra, he always put the money in those boxes. Sometimes it was as much as a dollar at a time, but mostly it was a dime or a quarter. Drop by drop, the bucket will always fill, Lev had told him once, and Jack repeated that like a mantra as every time he walked across the shiny marble floors of the bank.

The reporters were out on the scene at the museum, and wherever there were reporters there were people hocking red hots. The price was higher this far north of the Lower East Side. He waited in line with Rosie and Chase behind a young reporter with thick mutton chops and a fashionable piercing over his left eyebrow.

"What do you think about your competition?" the reporter asked. "Professor Watkin's steam-powered Scourer?"

"If it was going to work they wouldn't have us here."

"They say they already tested it at Princeton and it cleaned six chimneys in an hour."

"They say all sorts of things."

"Can I rub your button?" The reporter didn't wait for an answer, and he pulled one so hard it came off in his hand. He didn't tip. Jack walked off and went up the pulley.

He was dismayed to see a number of reporters on the roof. There were fewer than on the sidewalk, but still enough to be distracting. He found Rosie waiting for him by the pulley operator and Chase by Dr. Watkin's brass contraption.

Watkins was pointing out various features to the gathered reporters with a reed cane while the museum's director just smiled. Two men in spotless overalls were making adjustments to the machine. Jack thought it looked like a shiny python. It was reticulated and had rubber tubes attaching it to a squat steel boiler. One of the technicians turned a valve and the machine squirmed to life, nearly knocking one reporter off the roof. The director never stopped smiling.

Two hours later, the reporters were sharing a flask and smoking cigars while the technicians continued to make adjustments to the tail of the machine. The director had lost his smile, and Jack knew it was time to go to work.

The Princeton inventor and his team were trying to control the steam leaking from the thrashing mechanical snake and still refused to give up hope.

"Perhaps you can start on the far side of the roof while the gentlemen get their machine tuned up," the director said, looking at his gold pocket watch.

Jack did not like top down jobs. Most sweeps prefer to enter through the fireplace and work their way up, getting more comfortable with the chute as they climbed higher. Every chimney was unique and had its own sort of complications. Jack mostly worked in the sorts of neighborhoods where most of his jobs were up-climbs, but in fancy areas owners didn't want grimy children trudging across their Persian rugs. They preferred to keep them out of sight and Jack understood why. Even in January he refused to use the fireplace in his rented room because he knew every shovel-full of coal was coating the bricks with creosote that a child would have to scrape off. He couldn't bring himself to do it. He just bought more blankets and ate his meals from carts.

Rosie was new, so she would go first—the top of the chimney had less soot and creosote. The creosote, also called gunk, was a nasty tar-like substance, the byproduct of wood and coal. It coated a chimney with a hard cake and would eventually set aflame if not removed. Gunk was filled with pollutants that lodged under the skin or mixed with sweat and caused cancerous tumors called soot warts. To remove gunk required a very sharp, broad chisel and strength. Since the chimneys in New York by law were only allowed to be nine inches square, the size of a dinner plate, there was very little room to get leverage. Rosie was too small to scrape thick gunk off the porous bricks.

Rosie went in, working her way down to the first elbow in the chimney. Most chimneys had at least three right angles to them, but some had many more. Had Rosie slipped, she would have only fallen to the first elbow, which would hurt but wouldn't kill her, and Chase would have easily been able to retrieve her. Jack had been working with her to help her get better at collecting the soot in her coat pockets. The wide wire brushes were hard for her to handle, so most of the soot spilled on her. The climbing coats had four oversized pockets for soot, and if they were filled they could bring in two cents each.

Jack helped Rosie out of the chimney.

"Alright Chase, gear up. Rosie, stop breathing until I wipe your nose and mouth," Jack said.

Jack started to clean the loose soot from her nostrils and lips. Jack emptied the soot into a pail.

"You got three pocketfuls this time. That's great. How much is that going into your box, Rosie?" Jack asked.

"Four?"

"No. Two, plus two, plus two. Work it out."

After a moment, she said six triumphantly.

Jack watched as Chase rubbed pickle brine on his knees and his palms. The vinegar hardened the skin and had some antiseptic properties.

"Where did you get those?" Jack asked, looking at the goggles Chase had put on.

"Mine got busted on the last job," Chase said.

"Okay. Remember left, right, and down to the grate. Watch out for the corner pieces. Take your time," Jack said, fitting the mask tightly around Chase's face.

"Back in a jiff," Chase said.

Jack wondered, as he watched the boy slide into the chimney's crown, if he would be able to fill the boy's bank box before he was too big to fit anymore. Some of the other master sweeps purposefully kept their pupils underfed, to keep them small. "I'll put some extra in his box tonight," Jack thought. His thoughts were broken by the sound of Watkins' machine blowing a gasket and the shouts of the workmen.

Jack checked the lamp—it was over half full, more than enough for this chimney—and the knot on the snuff line. He rubbed the reflecting mirror and gave it to Rosie.

"Listen for Chase's instructions. If you need to yank it up because of fire; pull this string and it will close the damper."

Jack checked his watch.

"Excuse me, my dear good man," Watkins said, approaching with a dropping Calabash pipe held between his teeth, "do you have a match?"

Jack pulled out his own chipped clay pipe.

"Where is that little angel from?" Watkins asked.

"Rosie? Greenwich Village, I imagine. I fished her out of Ogsby Orphanage."

"How old are they?" the academic asked.

"Rosie is probably eight. Chase is eleven."

Jack grew uncomfortable with the conversation. The professor was polite, but he could see the judgment behind the lenses of his spectacles. Jack was used to other master sweeps judging him for being too soft with his wards. But he didn't need this professor from Princeton judging him. He was capable of doing that himself. He had known exactly who he was since that night in Tennessee when he dropped his government-issued rifle and ran. What he did was not atonement in the way Lev would understand it, but simply a way to get through the day.

"What are you a professor of?" Jack asked changing the topic.

"Philosophy, actually. I am the Emerson Chair to be exact. I am not an engineer and really have no head for machines. I spent not an inconsiderable amount to bring in those two Swiss lads to design this device. They are amazing technicians, but the problem is quite confounding. It worked perfectly in Princeton and for the life of me I can't figure out what the problem is here… it is quite frustrating… especially with the newspaper men here."

"What chimney did you test that thing on," Jack asked.

"The main boiler for the campus. It was twice as tall as this one and it worked beautifully. Not a hitch."

"That's because it was a straight chimney, I would bet. Like factory stacks. None of the twists and turns. That's where your thing is getting caught up. It's too big."

"Nonsense. It is not even seven inches in diameter. That is not the problem…"

"Not too thick, too long. It can't make the two elbows after the first damper. Probably gets caught up on the second corner piece. It angles like this," Jack said, bending his fingers downwards.

"We had sweeps consult with us. Told us what brushes to use and everything, but they never…"

"Cause they don't know spit. No sweep does. Only the yipsels know. No sweep has even been inside. I only know what I know because Lev, a friend, designed chimneys. You need a mason and not some two-bit sweep if you ever hope to get that contraption to work in anything but stacks."

"I guess you have a good night's work then. Thank you. You have been very informative."

The professor started back to his mechanics but stopped.

"I must be imprudent and ask you something."

Jack just shrugged.

"You are obviously a man with not a small spark of intelligence. Why would you work such a loathsome job? You must know what you do, what those children are put through. It's frightful. Just last week in Brooklyn two children died in a chimney fire. The owners of the house hadn't even properly extinguished the flames in the fireplace while they worked. They may be black, but they are still God's children. It is a moral outrage."

"And who cleans your chimney, Professor?"

Before the good professor could give a response, Rosie called out. She called the boy's name over and over, her voice getting louder and the pitch more desperate.

"I heard him cry. There was a sound. I heard him cry," the distraught girl said, starting down the crown.

"Did you damp the lantern?"

"Yes."

"Pull it up."

Jack started to call for the boy but there was only the echo of his voice rebounding on the bricks.

"You need to go down. Forget the tool belt, just find Chase."

Jack helped the girl tighten her goggles. She started in the chimney, putting her knees and elbows at a diagonal to the brick surface.

"If you smell smoke call out to me. You will doing it in the dark, can't risk the lantern."

Jack could see the whites of her eyes through her smudged goggles.

"Nothing to be afraid of in the dark. I'll be right here."

Forty minutes crawled by before Rosie popped her head back through the crown. She pulled the cloth from her face and breathed deeply. She didn't need to say anything; Jack could smell the smoke on her braids.

"He's alive. I touched him. He's stuck, knee to chin. There was fire in some gunk he was working. He tried to get out fast and got stuck at an elbow. I couldn't move him. You got to save him."

Jack had always warned the children to touch a creosote deposit with their palm to see if it was hot before chiseling away at it. Gunk could smolder for weeks and flame up when exposed to fresh air.

"He can't talk. He is suffocating!" Rosie screamed, throwing her goggles.

Jack knew that it was not uncommon for new yipsels to misstep in their climbing, getting their knees stuck by their chin and thus wedging themselves in the shaft. He had schooled his wards that if this were to happen they needed to relax and not panic. Panic caused the lungs to expand, thus getting them more stuck. The key was to relax and let the muscles uncramp. This was of course easier said than done, particularly when there was a fire underneath you. Jack was grateful he'd spent the extra money to outfit his yipsels with leather clothing that was more flame retardant than the undertaker's castoffs. But still, Jack knew there was nothing he could do.

"You have to save my brother," Rosie cried.

Brother? Jack hadn't had any idea they were siblings. It now made sense. Chase had insisted Jack go to the Ogsby orphanage when Malcolm left. Rosie came right up to Jack and asked to be taken on as a yipsel while all the other children tried to avoid eye contact with the strangely-dressed man. Chase went through a routine to check her leg strength and look at her hands and gave her his approval. It had all been a show for Jack, but it had worked. He now realized that he knew almost nothing about these yipsels. He provided the best he could for them, put money in their deposit boxes and hoped for some vague, undefined better future for them but knew nothing about them. If he did, he wasn't sure he could have sent them into the chutes.

Jack jumped down from the ladder and nearly collided with the professor.

"What is the problem?" the professor asked.

"The boy is stuck and there is probably a fire. I need to get some lard from the red hot hockers."

The professor called out in German to the workers who were still tinkering with the chimney machine. In a few minutes the pulley operator was bringing over a pail of lard from the one of the vendors. Jack had already stripped to his underwear. He was slightly embarrassed by the reporters' and the museum director's stares.

"Hurry. Help this man," the professor called, pulling off his kid gloves and plunging his hands into the sausage grease.

Jack felt warmed by the grease as he climbed the ladder. The professor handed him the lantern.

"There might be a fire. Can't chance it."

"You need light," Watkins said. "Wait just one more moment."

Jack sat on the crown, impatiently waiting. He could see one of the workers kneeling by the machine with the professor gesturing towards him. They returned with a piece of the machine.

"It's the tail light. So we can see the machine work. It has a lead battery. It is heavy, but small enough to fit on your head. It should last thirty minutes."

"You will need to lower me," Jack said. There wouldn't be room for him to climb or even move.

Jack tied the lantern rope around his ankle. He plunged his head in while someone took his legs and held them up, pointing him down into the chimney. He positioned his hands like a diver.

Jack could see little except the brick, the brick that scraped his shoulders raw in the first minute of his descent. The pain in his shoulders was soon equalled by the rope cutting into the flesh above his ankles. He felt like he was being digested by some stone monster.

His hands felt the first elbow that turned sharply to the left. His fingers searched for something to grab on to. At long last he found crack between two bricks. He pulled himself inch by inch around the elbow, each vertebrae in his back clicking over the edge of the cornerstone. It was slow and painful to move horizontally. He had more control, but it was hard to squeeze through and there wasn't even room enough to turn his head. The tip of his nose rubbed the brick as he wormed his way through the twenty feet of chute to the next elbow.

This elbow went down. He was struck by a rising panic. The light of his lamp bounced over a swallowing throat of blackened brick. The muscles in his arm ached. He knew there would be no way to reverse his position, and no way for those on the roof to pull him past the elbow. He was alone and trapped. It would take days for the masons to remove the wall and find his dead body. The fear in him forced his chest to try to expand but there was no room. He closed his eyes. *Keep moving*, he told himself, and then pulled against the elbow forcing himself to drop head long into the next chute. He only slid for a few feet before the rope tightened; threatening to slice his feet off. Slowly he started to descend. He relaxed his arms and let gravity do the work.

Jack first saw the wire brush.

"Brush," he said. "Move it."

Jack put his palms against the brick to stop his descent. There was no way he could get around the brush. He enjoyed the lack of pressure on his ankles as the rope started to slack because he was holding himself upside down with his palms and his thighs.

"Got to move brush," Jack whispered trying to keep his mouth shut.

Jack watched the soot dance around the bubble lenses of his goggles by the dim light of the headlamp. Then slowly the brush started to move. It moved down a few inches and stopped.

"Can't. No room," Chase whispered.

Jack tried to reach out to the brush with one hand but started to slip down. Jack counted to three, hoping there was enough slack in the rope. He pulled his hands away from the brick and let himself fall.

Jack felt the brush slash across his scalp as he crashed into the trapped boy. They fell for a few moments in darkness. Then there was a bone shaking crash.

Jack realized they were just above the fireplace, lying on a wrought iron gate. Jack worked to loosen the knot at his ankles but gave up and removed the chisel from Chase's belt and severed the rope. They had fallen twelve feet onto the grate and were bruised, but nothing was broken. Jack inspected Chase and found his hands blistered.

"There's a gunk fire," Chase said.

Jack realized they must have fallen through the fire zone on their way to the gate. They were trapped between the chimney fire and the grate.

"We'll never make it back up. I doubt they'll hear us from down here."

All they could hear was the sizzling gunk above them.

"We probably cracked it open further," Jack said, listening to the flame. "As more air comes down the chimney the fire will grow."

As Jack was saying this, burning stalagmites of creosote tar fell to the grate. Chase quickly brushed flaming tar through the gate where it fell on the logs in the fire box.

"More will fall. Don't let it get on your skin. It will set the logs on fire. We have no choice but to put it in the firebox."

"Then we'll burn," Chase cried.

"The smoke will kill us first, probably," Jack said, more to himself than to his young charge.

After a few more burning gunk blobs fell Jack grabbed Chase's tool belt.

"We have to get through this grate. Then we'll drop down and be out. We can then go up top and cap the chimney to smother the fire. Alright?"

The grate was well secured. Many rich homes and businesses had ember grates installed right above the fireplace to protect the places from snake men, who would send children through chimneys to steal valuables. The area on the grate was wide, nearly four feet by five feet, but only about eighteen inches high. Jack was thin but tall so he had to lie down nearly flat to chisel the mortar.

Finally it started to give.

"Kick it, Chase. Put your back to the wall and kick it hard, right here," Jack said.

The boy gave it three stout kicks and the gate crashed on the smoldering logs. Jack and Chase went sprawling onto the marble of the exhibit room.

Alerted by the noise, an elderly security guard burst in the room with his gun out. Chase was standing up, trying to help

Jack disentangle himself from the fireplace screen. The light from Jack's headlamp flashed in the guard's eyes and the man pulled the trigger. Jack heard Chase scream.

Jack didn't look at the director but stared at the chain-smoking guard in the corner. The man was as old as Lev and seemed unable to keep his hands from shaking. The director started to count out bills in his hand.

"This is most unfortunate. We are willing to pay, of course. The negro will be taken out in the morning. No reason to make a fuss in front of the reporters. That would not do anyone any good. No sir, not anyone. Terrible accident. A fluke, really. We have a new exhibit and don't need any bad press. So here is a bonus. I think you will find it very generous Mister…?"

"Candlestick," Jack said, taking the money.

"Mr. Candlestick. I trust you will keep your own counsel about these things. I will let you out the side."

Jack started to walk out when the director reached over and touched Jack's button.

"We can all use a bit of luck; don't you think Mr. Candlestick?"

Calamity set the paper down and picked up his coffee. The morning light was trying to make it through the kitchen's window.

"It seems we will need that laudanum for more soot wart medication after all," Calamity said to Pip. "The esteemed Dr. Watkins' machine has failed to clean the Met's pipes."

"A setback," Pip responded. "The reformers aren't going to quit. There will be another competition. Someone will invent something as long as there's money to be made."

"Those damned reformers don't need money. They got all the money they need. They just want to feel less guilty, that's all," Mathilda said, munching on some toast.

"Well, I think it is a good thing," Pip said. "I hope they find a machine that can do it. It's the devil's work going down those smoke-stacks. Those kids deserve better."

"You really think it'll matter, my dear boy?" Calamity said. "Even if our great woolly minds come up with some successful device, I have no doubt they will come up with some equally terrible jobs for children."

"They always do," Mathilda said, taking Calamity's coffee.

FREEDOM OF THE MARVELOUS

By Franklin Rosemont for the Catalogue of the World Surrealist Exhibition, 1976
Reprinted here by kind permission of Penelope Rosemont

Illustration by Allison M. Healy

CAUGHT UPON AN EMOTIONAL PRECIPICE BETWEEN THE IRRETRIEVABLE and the unhoped for, men and women today rarely recognize each other, or even themselves. Ask them who they are, what they are doing, where they are going; they stare blankly, stammer, look the other way. No one dares to be happy: too many wars, too many suicides, too many unemployed, too many priests, too many cops; too many "troubles" of every sort conceivable and too many that are scarcely conceivable at all. The exceptions prove the rule. The traffic is always heavy, the weather is always bad. No doubt about it: life today is only five percent of life, and day by day the percentage goes down.

Stand for twenty minutes on any street corner of any large city, and note the expressions in the eyes of the multitude of far-from-perfect strangers who pass by with nothing better to do, it seems, than to perfect their estrangement. At such moments people have been known to ask themselves: Is this why I was born? Is this all? They ask themselves, I say; for if they were to ask those questions of others they would find themselves placed under arrest, or carted away for electroshock, or provided with still further examples of the consequences of exercising the right of free speech.

It is nonetheless true that this discomfort, this despair that gnaws through the darkness in everyone's heart, endlessly opens loopholes in the walls of logic used to justify the universal immobilization of the human spirit. Through these loopholes, while they last, the flame of freedom faintly glimmers, illuminating the promise that something else exists or could exist. For most people this flame flickers its last, or almost its last, in a protracted form of death known as "maturation." Every hope is extinguished as fast as it appears, accompanied by rituals of submissive evasion that reinforce the cheap mysticism of everyday life. Imagination, dream, fantasy, play, adventure—everything that gives life a hint of magic and exaltation is relegated to an increasingly depreciated childhood. Civilization is founded on the murder of children because it is childhood, as Andre Breton suggested, that "comes closest to true life."

What remains, for most people, are only a few rare "unconnected" and "inexplicable" moments: fleeting eruptions of inspiration, sudden passions, dazzling encounters "by chance." Such moments, true glimpses of the Marvelous, secure themselves permanently in one's psychic life, in the depths of our inner mythology. Shunned by repressive reason, persecuted by routine, these magic moments nevertheless remain secret signposts for the wandering mind—for the shadow in search of its substance.

To extend these moments, to unite them, to hasten their proliferation, to arm them, as it were—such have always been central functions of poetry. With the formation of the surrealist movement the poetic effort has attained its highest stage of development. What had been only individual, sporadic, unconscious—and therefore easily defeated—with surrealism becomes collective, systematic, conscious, invincible.

To overcome the contradiction between these marvelous moments and the everyday, to actualize the Marvelous in everyday life—that is the surrealist project.

surrealistmovement-usa.org

By Miriam Roček
(AKA Steampunk Emma Goldman)
Illustration by Geneviève Lamarche Reid

A HEALTHY ALTERNATIVE

If you've been in the steampunk or dieselpunk subculture for any length of time, chances are you've encountered a few steamnazis. You know what I'm talking about. Their costumes range from full-on SS uniforms with goggles thrown on, to fascist-inspired, somewhat fetishistic uniforms with vaguely ominous made-up insignia. If you ask them about their costumes, they'll say that they're not into fascism, they just like the style. I don't like them and for some reason they seem to be a big thing in dieselpunk. The ones that are directly Nazi in their insignia make me very uncomfortable, while the more vaguely fascist-inspired ones just strike me as a little gross. Their reactions to being confronted with my, or others' discomfort, have ranged from vehement, vocal defense of their costumes to frank, apologetic, and respectful apologies and the removal of Nazi insignia, which is nice but makes me wonder why they put the stuff on in the first place. The basic consensus among these guys (they are mostly guys) seems to be that fascist-wear just looks cool, and that they don't much care what the implications of any of it might be. Whether the look is even all that great to begin with, which I think is pretty debatable.

What I'm going to do today is offer an alternative, and suggest some costume ideas (illustrated by the wonderful Genevieve Lamarche Reid) to show that there are many different, non-fascist, in fact, *anti*-fascist options for sources of inspiration in dieselpunk.

And now a disclaimer, because this topic has gotten me into arguments in the past: I am not telling anyone they can't wear what they want to wear. I clearly don't have that kind of power, and wouldn't use it if I did. Free speech is good. I am also not saying that everyone who fought against fascism during World War II was a saint, or that none of the people who did so committed war crimes. What I am saying is that I dislike seeing fascism romanticized in any way, and that I think anarchists and other anti-fascists are under-represented in a lot of what I've seen of dieselpunk, despite their being on the right side of history, not to mention badass as all Hell. I'm saying I'd like people to look at the rest of this article, and draw some inspiration from it for future dieselpunk costumes.

Lastly, I hope no one interprets me suggesting these people as inspiration for fiction or costumes as disrespectful or trivializing them, or their fight. Rather, I think that we draw on reality in our art, whether we're creating novels, music, or creative costumes, and that it's better when we draw on things that excite us. It's always good to think about the implications of the work you're creating, especially when it has roots in recent, violent history. (Remember, by the way: as always, when dressing in military uniform, that many people consider it bad form/offensive to wear real decorations you have not earned. And, as always, it's your choice whether you want to do something that might be seen as bad form/offensive.)

Mostly, I think I'm saying I'd like to see more anarchism influencing the fashion at steampunk events. But, I mean, come on, who wouldn't?

So. We all know a lot of people fought against various fascist armies in World War II. Instead of focusing on, say, putting on an American or a Russian uniform instead of a German one, let's look at the more interesting options. Interesting, not just in terms of costume, but in terms of the potential backstory you can have for your persona, if backstory and persona is a thing you do. (While I'm talking about costuming here, any one of these ideas could also be a character in an RPG, a story, a song, a play, whatever it is you do creatively.) The characters illustrated below could come from many different countries, and many different ideologies, just like the people who fought fascism in real life, who were incredibly diverse in terms of their ethnicities, geographical origins, and political and religious beliefs. What all that means for dieselpunks is that it should be possible to put together an interesting alternative history backstory for a character that incorporates whatever aspects of your own background you want to include, or whatever ideas you think might be interesting to play with.

One thing I love about anti-fascist fighters as a source for costume and character inspiration is the number of options they offer for female-presenting people and characters. This is because, on the whole, women had a much more active role in fighting against fascism than they did in fighting for it. This, in turn, is primarily because fascism sucks, and because a lot of the leftist groups fighting against it tended to have much more egalitarian approaches to gender. *Milicianas*, or female militiamembers, in particular became icons of the fight against fascism in the Spanish Civil War. Their images adorned recruitment posters and were featured in articles praising the fighters. These women were nine kinds of hardcore; they killed a ton of fascists while at the same time frequently being expected to do laundry, act as medics, and cook because, you know, woman. But in many cases they were treated as equals by their male comrades, and, notably, they wore the same uniforms.

TO FASCISM IN FASHION

This woman and her child are wearing costumes inspired by the Spanish Republican militias.

This one has lots of fun vintage fashion elements, while at the same time representing a group of people who are actually worth celebrating. Now, when I say that the child's costume was inspired by actual Spanish anti-fascist militias, I do *not* mean that it's a shrunk down version of something I found a picture of an adult wearing. That image is based on one of a Spanish child in the uniform of a local anarchist regiment, while the costume worn by his parent-or-guardian is likewise lifted more or less entirely from countless images of real-life women who took up arms alongside their male counterparts. (First child to approach me at a steampunk event dressed as a civil war-era Spanish anarchist wins the cookie of their choice.)

Now, this guy's outfit is based on various images of Italian, French, and Spanish partisans. He could be from anywhere; it's kind of up to you as you make up your character's backstory. There's a couple of things I really like about this costume. First, he's got his goggles on top of his hat. For some reason, this is a shockingly controversial choice in the steampunk world (maybe more controversial than fascist gear, in certain circles); a lot of people seem to feel like if goggles are up there they aren't easily accessible, which makes them not functional, which means they shouldn't be there at all. Now, leaving aside the issue of who decided fashion can't contain nonfunctional elements, since in this case, I'm pretty sure the guy is going to want every component of his equipment to be useful, I just want to go on record as saying that the people who say that are completely wrong. Wronger than wrong. And I will tell you why: I've seen tons of vintage photos of pilots, drivers, motorcyclists, welders, and others with their goggles pushed up on top of hats or helmets. They're out of the way, and apparently not too hard to get to if you need them. And, not trivially, they look really awesome. Which is, of course, a function all its own, though maybe not the one most on your mind when you've fascists to kill. This guy is ready, not only to kill fascists, but to sneer at anyone who tells him he's wearing his goggles wrong.

The other great thing about this guy's outfit is that it can be assembled entirely (apart from any sci-fi-ish additions you care to make) from stuff you'd probably be able to find in an army surplus store. Costuming on a budget, for the win!

(Except the gun. You probably shouldn't be bringing that to events anyway.)

This guy is ready, not only to kill fascists, but to sneer at anyone who tells him he's wearing his goggles wrong.

98 A Healthy Alternative to Fascism in Fashion

Of course, you *can* keep your German military gear, if you are for some reason wedded to it. In this outfit, based on a photo of a female member of the French Resistance, a dieselpunk resistance fighter wears a German military tunic, belted in at the waist, with the sleeves rolled up for action. It pairs well with a nice pleated skirt and a desire for liberty. Nothing puts the punk back in dieselpunk like repurposing military gear.

Waves of badass emanate off of this woman. Her civilian skirt and shoes are a good reminder that a lot of the people who fought for the resistance weren't trained soldiers, which means that you can create a character who was an actor, telegraph operator, veterinarian, or airship mechanic before they joined the resistance, and who uses the skills that they have however they can in their fight. And of course, that repurposed German tunic adds yet more richness to your backstory; your character may not be fighting for a country currently able to hand out medals, but her clothing shows you a victory she's won. Or rather, two: one, she very likely took an enemy's uniform from his corpse. Two, she looks way better in it than he ever did. Obviously. Oh, and another fun historical detail about it is her armband, which in the absence of uniforms, resistance fighters would often use to try to get the other side to recognize them as enemy combatants. A cool idea for alternate-history dieselpunk guerillas to draw from.

So, I realized I lied a few paragraphs back. I said that nothing puts the punk in dieselpunk like repurposing military gear, when in fact, nothing puts the punk in dieselpunk like my next topic for inspiration, which is the Battle of Cable Street. Holy crap, do I love talking about the Battle of Cable Street.

The Battle of Cable Street, for those who haven't heard of it, took place in London in 1936. It was brought on by the decision of British fascist Sir Oswald Mosely to lead a (legal, permitted) march of seven thousand uniformed members of the British Union of Fascists towards the East End, a poor neighborhood with a high Jewish population, not to mention a large number of working class political radicals. (There was probably a lot of overlap between those two groups.) Those who opposed the march, which included Jews, communists, anarchists, union members, and basically anyone who wasn't a complete asshole, got together to put up and defend barricades blocking the march route, and faced off with the London police force when the cops came in massive numbers to clear the road for the fascist marchers. Over three hundred thousand working class Londoners fought violently to hold the lines, even going so far as to kidnap police officers, which is just awesome, since as we all know cops do that to protestors all the time. A captured London bobby's helmet would make a great starting point for a costume, I feel. Just remember to raise your fist in the air and shout, as the Cable Street anti-fascists did, inspired by the Spanish anti-fascists, "no pasarán!" (It sounds great in a Cockney accent.)

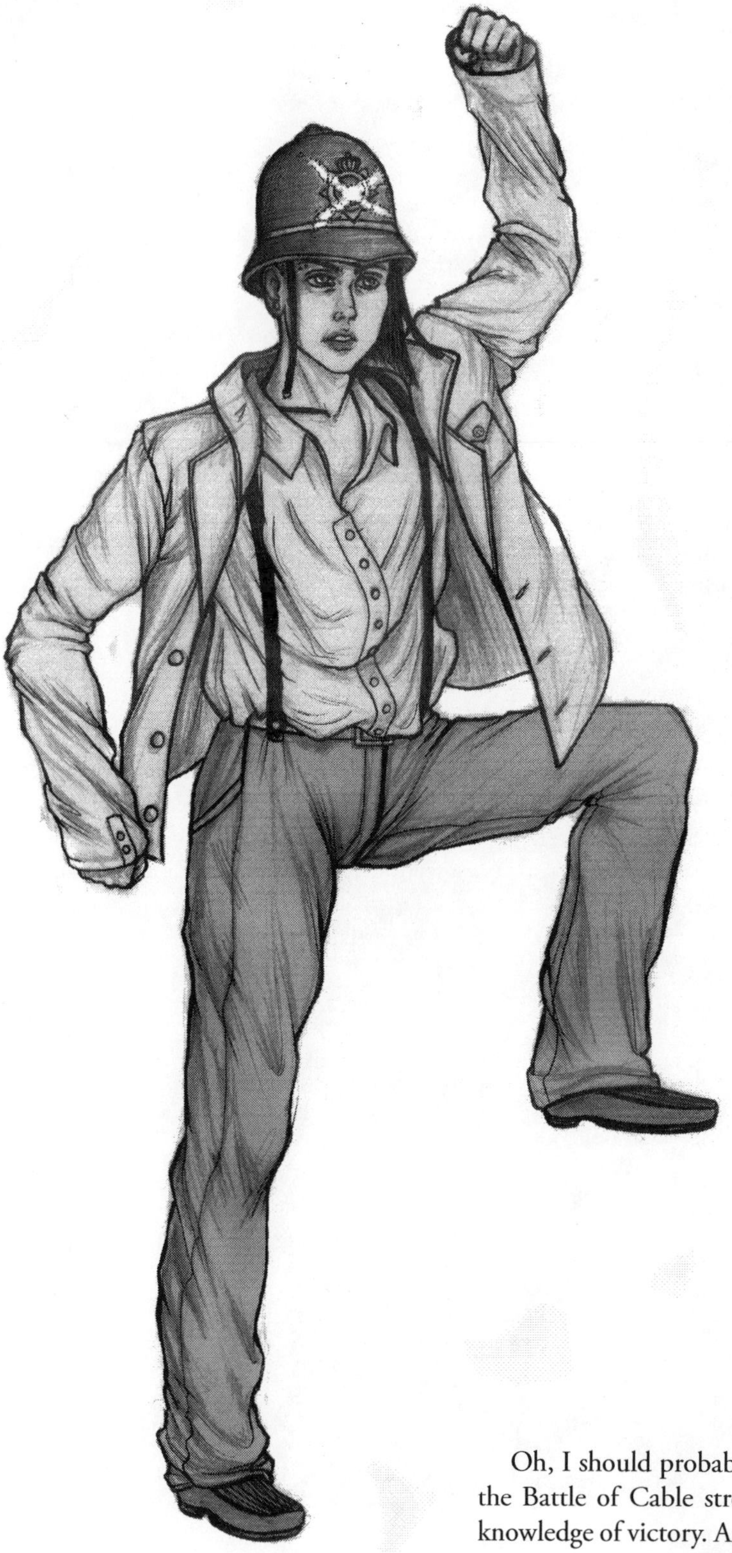

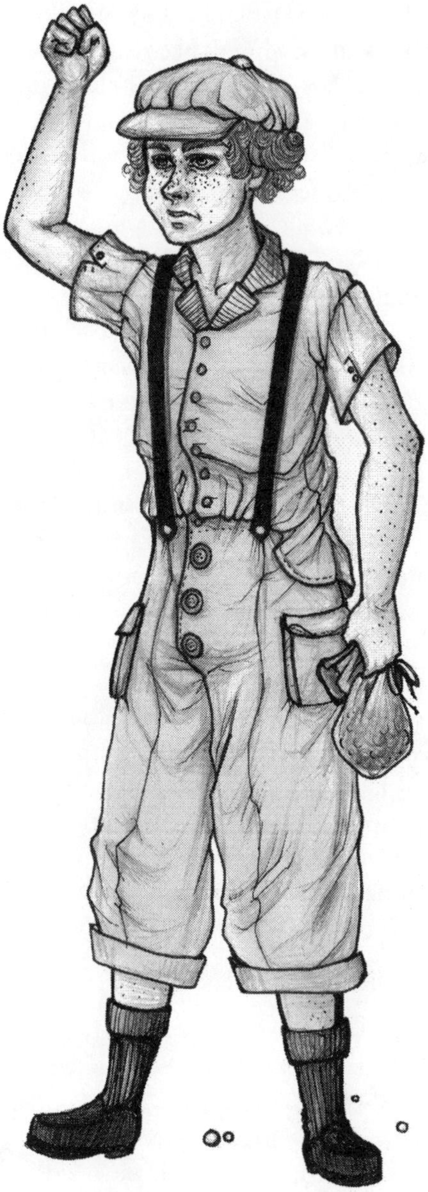

Oh, I should probably mention that the march was canceled, which means that the Battle of Cable street was a win. So you can wear your costume in the full knowledge of victory. And what could be more punk than a working class Londoner fighting cops in the name of stopping the literal march of fascism?

This is another one that comes in adult and child. According to people who were there, the fighters at the Battle of Cable Street included children who helped using a skill more or less unique to old time kids; they rolled marbles under the hooves of oncoming police horses.

This kid and that Spanish anarchist kid from the first image should definitely hang out.

What could be more punk than a working class Londoner fighting cops in the name of stopping the literal march of fascism?

Our last costume/character idea is the only one to be inspired directly by an individual historical figure. Salaria Kea was a black nurse from Ohio. She was working at the Harlem Hospital when she joined the Communist Party in 1934, and in 1937, she joined the Abraham Lincoln Brigade, an American volunteer force that went to Spain to fight on the side of the Republicans. She worked in a hospital, treating wounded soldiers, one of whom she totally ended up marrying, because some people's life stories read like movies even before a screenwriter adapts them. The character in this image is inspired by a picture of her in her uniform. And yes, she really did have a cape. Because superheroes get capes, goddamnit.

So, hopefully those are a few images to get your imagination going for some ass-kicking, fascism-stomping dieselpunk stories and costumes. Incorporate whatever alternate history or character ideas you've got floating around. Should you meet a steamnazi while you're so dressed, maybe give them a history lesson. Or an alternate history lesson. And remember that not only are you dressed like way less of an asshole than they are, chances are you look better too.

Notes from the Bucket Shop

By Professor A. Calamity
Illustration by Benjamin Bagenski

Welcome, gentle readers, to the first column in a new series which will uncover some of the most closely-held secrets of the Bunkum Brotherhood. I seek to turn up the gas on the shady shenanigans of hucksters, swindlers, card sharps, and pub phantoms in order to educate and entertain the unwashed masses, and thusly have I scoured the Bowery's bucket shops, side-strapped saloons, suicide halls, opium dens, malicious midways, and fast-rolling traveling shows to bring you the best of 19th century chicanery.

We will start our sordid sojourn with cheats, then graduate to cons, and end with a classic swindle. Sharpen your nibs and open your notebooks—school is now in session.

Cheat: Rigging Dice

Dice rigging comes in three main varieties: weighted dice, shaved dice, and the tappers. Each has their advantages. Weighted dice are usually made with lead "slugs" that weigh less than an ounce. Slugs are added by boring out a pip (one of the dots on a die) and putting a drop of melted lead in the divot. Do not worry, the temperature required to melt lead is about 325 degrees, and bone, ivory, and even high-impact plastic dice only melt/burn at a much higher heats. Beware, though, lead fumes are nasty—so do it in a well-ventilated area. Make sure you paint over the slug with an enamel paint (easily obtained at hobby and art supply stores) the exact tint of the other pips. You want to put the slug in the opposite side of the die from the number you want the die to roll (for example, if you want the die to roll fives, plug the two). Weighted, or as they are often called, loaded, dice are no guarantee you will get the number you want every time. A good loaded die will roll the target number about half the time. To increase your odds, toss the die to achieve maximum number of rolls and minimize bouncing; this is best done by flipping your wrist and keeping your hand a few inches above the table. A little known fact is that early craps tables (and most modern ones) have a high wall around them called a "bump" that makes throwing loaded dice more difficult.

Dice with the sides shaved to manipulate the roll are called "trip" dice. It is much easier to shave bone or ivory dice than the Lucite dice commonly used today. If you are shaving Lucite dice, choose dark colors to minimize scratches (called "scars"). You can shave one side with a die cutter or with a very sharp razor and plenty of patience. The face you shave will be less likely to turn up than the other five faces, thus giving you a slight advantage. A well-shaved die is very hard to detect and casinos have used laser micrometers to determine the fairness of dice. Trip dice are a lot less useful than loaded dice because the odds of getting a crooked roll is much smaller and it only limits a certain number from coming up rather than forcing a face like a plugged die. Shaved dice also wear out more quickly than regular dice, depending on the surface.

The Holy Grail of rigged dice are called tappers. Tappers were invented sometime in the early 19th century and are extremely hard to find, even on the internet, often costing hundreds of dollars for a single die. Plastic, non-translucent dice (like a standard white die) are the easiest to transform into tappers. Tappers have small interconnected tubes drilled precisely inside of them. The end of the tubes near each face (as opposed to the center of the die) is slightly larger and this slight reservoir is called a "dimple." Before the holes are sealed, a drop of mercury is added to the tunnel. This very heavy and nearly frictionless liquid will come to rest in a reserve, creating a loaded die effect. The key to using a tapper is to tap the die on the opposite face of the number you want to come up (example if you want a 6, tap on the face with a 1). The tapper allows you to pick which number you want to favor with each roll, while giving no advantage to the other player shooting with the same die (unlike the two previous dice cheats). Please remember mercury is toxic and care should be used.

Con: Shell Game

Now we move on to a con where you always win. It is an ancient con, but it was very popular with the riverboat and train crowds in the 19th century and you can still find this pigeon pluck on the streets of New York, Paris, and other urban areas. It is the classic shell and pea game. One of the reasons this con is so popular is that it is fast and the setup (called a "stint" by Victorian bunco-artists) can be carried easily in your pocket. In later columns I will go into how to rope in a mark (sometimes called a fish), build a bustle, and swing a fob—all essentials in achieving maximum effect/profit from street cons like the shell and pea. For now we will focus on the execution of the mechanics of this con.

It is important to use a walnut shell (plastic and even brass, for steampunks, walnut replicas can be easily purchased at a magic supply store) because of its peculiar properties. Walnuts are quite heavy and hang from trees and this requires a substantial stem. This botany lesson is important, and I beg the reader to exercise patience. The stem creates a slightly puckered end, referred to as a "bung." The bung is slightly raised in the back and the front is rounded. The outside of the shell is ribbed, allowing for maximum grip with the fingertips.

A pea or small piece of green rubber or foam is concealed underneath one of the three shells. The shells are shuffled back and forth while the mechanic rattles off their patter. The mark follows the shell with the pea with their greedy eyes while putting a wager on the table/stand. Always make sure the mark puts their coins, bills, or doubloons actually out on the table, as otherwise it is very difficult to get paid by the soon-to-be sore loser. The mark points to a shell and it is lifted, revealing nothing underneath. You quickly snatch the money and leave. So how is this done, you ask?

I trust you remember the bung from the earlier discussion; this is the key to the whole deception. When a shell is pushed straight forward, as opposed from side to side or at an angle, the pea can easily pop out from the bung and be concealed by pinching the thumb and the third finger together. To make it look natural, place the index and middle finger on top of the shell pointing towards the mark and the thumb and third finger pushing the shell from behind. The fact that the index and middle fingers can easily force the shell in any direction because of the shell's ribs makes pushing with the remaining two fingers unnecessary. It will look natural if you always do it this way. Once the pea is safely concealed by the thumb and third finger, it can be "loaded" into another shell simply by dropping the pea directly behind the shell and dragging the walnut back over it. The pea will quietly and secretly slide back through the bung. To be safe, you should never reload a shell until after the mark has chosen their shell—that way you can't lose.

Some key details need to be pointed out to make this con a success: line of sight is crucial, so make sure the mark doesn't have too much of a bird's eye view of the playing surface. Play at a table sitting down or at a stand that reaches at least to your solar plexus. While shuffling the shells, periodically lift the shell to show everyone the pea is still there. Never move a shell directly forward with a pea in it during the shuffle because the pea may slip out. Certain very smooth surfaces like glass tabletops and well-polished bars may make the pea too difficult to get out, while felt and cardboard work very well. Never let the mark be the one to flip a shell because they may flip all three before you have a chance to reload. Let the mark examine the shells and the pea if they wish, but never let them shuffle a shell with a pea in it, for it might pop out the bung. A good patter and an accomplice that "wins" in front of your crowd are key to selling this con to a fish. Have your accomplice select the correct shell after the mark has selected the obvious, but false, one.

Swindle: Glim-Dropper

The Glim-Dropper swindle dates to the mid-19th century, and it was made popular by a riverboat grifter named George Duval. This bunkum requires two hucksters and a greedy mark. Luckily, one is born every minute. The actual *glim* (meaning something shiny, something that glimmers) could be any number of things: a glass eye, a wedding ring, a safety deposit box key, or a pocket watch, to name a few examples. For some reason the glass eye was the most popular. This trick was most often performed in a saloon or an expensive restaurant. A patron would suddenly start searching for something on the floor. She would claim to have lost something valuable that has sentimental value to her. This glim would be searched for in vain by patrons and possibly the owner. The patron leaves her address (phony of course) and offers a thousand dollar reward for the object's return.

The next day the fellow huckster enters the same establishment and orders a beer. He sets down a glass eye (or whatever the glim is) and starts fiddling with it. He claims to have found it right outside the premises. This huckster looks poor and down on their luck and asks the owner for a nearby pawnshop to sell the glim. The huckster explains he expects to get a hundred fish backs for it. If the owner is honest, he will give the poor patron the address of the woman who lost it so he can collect the reward. The con depends on the inherent dishonesty of people, which often pays off. If the owner wants to get the reward themselves, they offer to buy the item from the poor patron for less than the reward. Reluctantly, the patron sells the actually-worthless glim. Needless to say, the woman is not at the address nor does she ever return to the establishment to give the reward for the glim. The owner is stuck with a cheap glass eye.

Next time we will delve into the 19th century secrets of why old west gamblers wore blue sunglasses and multi-colored vests, the classic con of the pigeon drop, and how to win a coin toss game called Downy Fields.

Disclaimer: This column is for entertainment purposes only, not to cheat your friends, loved ones, or annoying fat cats with too much money in their pockets. Trying these swindles will likely put you in conflict with flatfoots and local magistrates. Don't use these techniques to give up your day job. Of course, if you don't have a day job...

THREE SIMPLE MEN

AN INTERVIEW WITH BB BLACKDOG

Illustration by Sarah Dungan

STEAMPUNK MAGAZINE: Who is BB BlackDog, for our readers who aren't yet familiar with your music?

DALE ROWLES: We're two bass players and a drummer, two English and a German, with a love of the steampunk style and philosophy, just trying to make some original honest music. What you get on the recordings is what you get live: we don't overdub, use click tracks, or autotune—we just mic everything up and play. Of course there's delay and reverb etc. on the recordings, but nothing we don't do live.

Three simple men playing simple and hopefully entertaining/humorous music. (We do have dancers of the burlesque and belly dance nature with us at many of our shows, but nothing too graphic, just suggestive—we are steampunks after all!)

SPM: *Your music reminds me of equal parts goth rock (Cop Shoot Cop, Big Black, even Danzig) and bar rock. How did you get your start, and who influences you? Sorry to ask such a constant-ask of a question, but I'm genuinely curious about your sound.*

DALE: Well, not an easy question really. BB BlackDog formed completely by accident, as with most things that happen to us—I was selling musical instruments for a living, and met a guitar builder (Stefan Becker) at a party in Germany. He had been given some studio time by a friend for his birthday, and said it would be funny to try and make some songs with just two bass guitars and drums, so we found a session drummer and wrote six songs in three days and recorded them in two. I thought nothing more of it until

I got a phone call from Germany saying they had been put on the net, that people loved them, and that we'd better find a full-time drummer and record more. So enter "Axel Boldt" (Chief Engineer Axel Boldt), a 6'9" giant of a German drummer.

We went back to the studio, recorded two albums worth of material, and booked our first tour back in June 2007.

Stefan left due to other commitments in 2010 and was replaced by John Ferguson (Baron von Gimphausen).

We never actually set out to be anything, just to write and play songs we like. We were adopted by the steampunk community, mostly after John joined, due to our stage outfits. But since then we've embraced it fully, as it just suits us perfectly.

As for influences, we're quite varied: Axel likes funk, Prince, grand master Flash, and the like; John is a big Pink Floyd and jazz fan. As for my influences, I grew up in the 70s so I progressed from T Rex and Bowie into Black Sabbath, Deep Purple, the psychedelic rock scene, and then Punk. But I'm quite varied in my listening now. I suppose it all comes through.

SPM: *Lyrically, you're quite a bit more down to earth (if you'll pardon the airship pun) than most steampunk bands. What drew you all to steampunk? What do you feel like steampunk has to offer?*

DALE: You're right—I actually try and stay away from writing about cogs, airships, and the like. I think life, humour, and love are universal. As I said before we really just fell into steampunk by accident and found a home. We'd been playing a few years and gigged a lot, but nobody really knew where to put us. We can be quite varied with our songs, from ballads to almost metal, funk, blues to punk, but because of our two bass sound, you can always tell it's us. The steampunks seem to like the variety and aren't trying to put us in a box, which gives us a lot of freedom. That's one of the things I really like about the scene—there's a lot of freedom to express yourself, and they're not a judgemental lot.

I hope the steampunk philosophy—of helping others, self-expression, understanding, quality, and honesty—spreads. It would be a fitting legacy for the most friendly of movements.

> That's one of the things I really like about [steampunk]—there's a lot of freedom to express yourself, and they're not a judgemental lot.

SPM: *Your song "Politicians" starts with the line "politicians, we hate you all," which is a nice, nearly-universal sentiment that I can certainly get behind. Do you all have particular political leanings, as people or as a band? I don't really care who you vote for, mind you, but I'm curious to know your thoughts more broadly.*

DALE: As you've probably guessed, our songs have a lot of humour. We don't really support or criticise any political leanings: there's good and bad in all. But as "look after yourself" seems to be the world's mantra, it's bound to be the same in the corridors of power and I just light-heartedly try and point out the obvious. I am a member of the "Official Monster Raving Looney party" Britain's third oldest political group, they pretty much have a similar philosophy, pointing out some of the absurdities carried on by all parties.

A lot of our songs just relate to experiences, re-state simple naive common sense, or are played straight for laughs. We've had a few people "not get it," but I really don't keep a gimp in my basement.

SPM: *From what I can tell, you all aren't geographically based anywhere in particular, with some of your band being from England and some from Germany? For those of us in the US and elsewhere, what can you tell us about the European steampunk scene? Is it fairly international, or are individual countries pretty insular?*

DALE: There's a large and rapidly growing scene in the UK, with groups all over arranging events and meetings. The National Festival at the Asylum in Lincoln is nearly sold out for next year, and a new summer festival starts in 2012. We're doing a full 3 hour special on the Friday night, with Abney Park flying over to headline the Saturday, so it's getting very large.

The European mainland has many solid little groups in Belgium, Holland, Italy, Austria, and Germany as well as the Scandinavian countries, the old eastern countries, and a few more, but it's still blossoming. There are quite a few people working hard, ourselves included, to start a European festival. Axel jokes that if there's something good going on in the US, the English will latch on and be not far behind, and vice-versa, and then either way Germany will be two years behind, and in the northern farming/fishing area he's from it will be a year after that!.

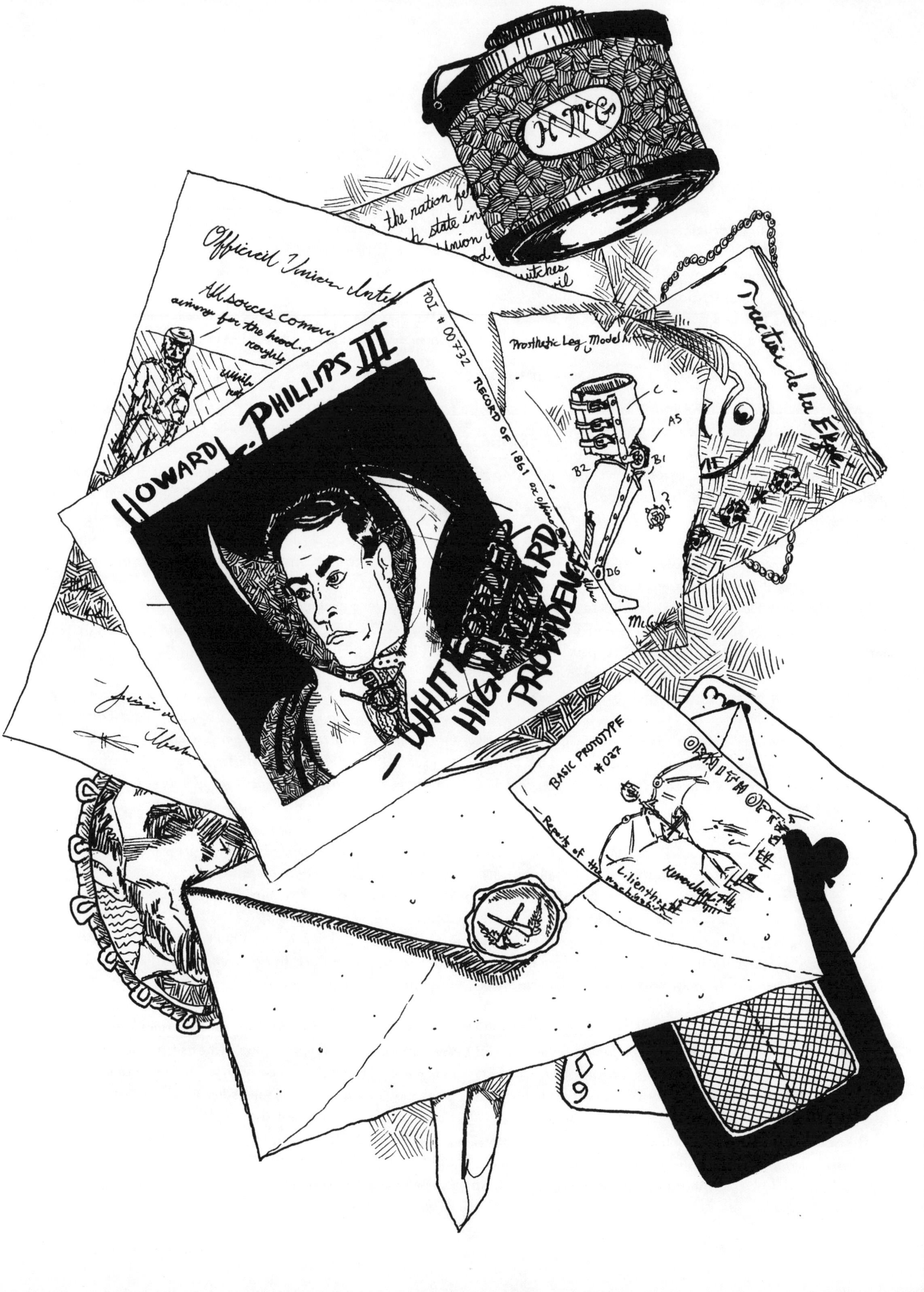
HOWARD L. PHILLIPS III
POI # 00732
RECORD OF 1861
PROVIDENCE
Prosthetic Leg Model
BASIC PROTOTYPE
#027
Lilienthal
H McG

The Scouts of the Pyre, Part 2:

The White Witches of Providence

David Z. Morris
illustrated by Benjamin Bagenski

"I DO SO UTTERLY LOATHE THE FOREST."

The man was spectrally tall and gaunt, with a nose whose vault could have supported a bridge. His black hair swept severely back from his high forehead and a stark white robe hung from his angular shoulders. His bone-pale skin added to the funereal air of his every word and motion. He gazed slowly around the clearing that housed General Grant's camp, his height lifting his gaze above the heads of almost all the bustling men, revealing to him the light and greenery around the edges.

"Nothing here but the rotting offal of trees, and the vermin who call it home." He gave a pronounced and dismissive sniff, and his gaze turned, ever so slowly, to meet that of Lieutenant Harry McGee. "The underfoot creatures of secret and lie."

General Ulysses S. Grant stepped forward with a slight cough. "Lieutenant McGee, May I present Howard L. Phillips the Third, High Wizard of the White Order." Grant seemed uncharacteristically infected by the witch's pomposity—or maybe just his height—and made a slight gesture, almost as if he were a servant presenting a prince or a vizier.

"Howard L. Phillips the Third of Providence, if you please," the witch corrected, not in any way acknowledging one of the rising stars of the Federal military. He made a flourishing production of presenting his hand to Harry, its fingertips dangling foppishly. Harry made some pains to enact a handshake, but in the end had no recourse but to limply grasp those dangling fingers with matching unmanliness.

"A pleasure," he replied, already sensing it would be made a lie.

General Grant harrumphed again, ever so demurely. "The High Wizard and his company were travelling with us when you arrived, and he has offered his personal assistance in pursuit of our stolen weapon. The mystic arts wielded by he and his cohort will be invaluable in your coming battle."

Phillips' gaze had not wavered from McGee's face. "You are not expert in the realm of the supernatural?" He gave a brief glance down at McGee's new peg-foot.

Harry was certainly no stranger to the disdain of his supposed betters, but something about this Phillips rubbed him particularly wrong. There was the matter of strangeness—the garb of a High Wizard was outlandish, the bull-horns of the starched collar rising some inches higher even than Phillips' head, bright rings of red encircling the waist and shoulders, the hem of the long robe incongruously stained with grass and mud. Then again, Harry was not unfamiliar with the eccentricity of the fey classes, coming as he did from a university town only slightly less notorious than Providence for intellectual and spiritual folly.

The High Wizard's companions, though, were a real source of unease. A dozen lower-ranking witches huddled in the background of the exchange, uniformly hunched, draped in long white sheets with hoods that hid their faces, and streaked all over not in natural dirt but in odd yellows and purples and dull reds that seemed too degraded to be decorative. They were uncannily silent, these witches of the rank and file, not speaking or making any other sounds of communication or effort. Whether through strange empathy or absolute authority, Phillips alone spoke for them, without hesitation.

"I certainly do greatly appreciate your generosity," Harry nodded. "I hope we are not taking you away from important work."

Phillips lifted his eyes and again regarded the near distance.

"Oh, all of our work is vital," he intoned with palpable disinterest. "Even this little errand."

But no, no. It was neither the silly clothes nor the strange bedfellows, McGee finally concluded. What he didn't like about Phillips was that he was an aristocratic twat.

Harry certainly did not think of himself as common chattel, but this Phillips, with his painfully affected faux-British lilt and blunt insistence on titles, was so patently enamored with the happenstance of his own high birth that Harry could hardly suppress his loathing.

"Well then, good fortune for all of us," he replied, tight-lipped but polite. "It is true I'm less adroit with a spell than a Springfield." He tried not to prematurely bask in his own magnanimity.

Phillips did not reply, but instead stood regarding McGee with insolent frankness. Phillips' brow creased slightly, and McGee saw a strange intensity to the examination. After silence hung in the air for a beat too long, just before McGee broke off his own gaze out of embarrassment, the General interjected. "It is imperative, I hope I have impressed on both of you, that you move with speed and do not flag. You must also move carefully, as this is of course a path through enemy territory."

Phillips' gaze mercifully left McGee for Grant. "Believe me, General, I share your sense of urgency. What we chase is a threat to all of us. I and my allies can assist with both speed and secrecy."

"Fine, excellent," nodded the General—just a tad more assured now that the talk had turned to strategy.

"What other arms can you spare, sir?" McGee took a risk in seeming too demanding of Grant, but he didn't relish travelling alone with this crew of witches—nor, frankly, fighting alongside them.

"As a covert mission, numbers would not be to your absolute advantage," the general replied. "And of course we have our own dire battles to fight with the armies here. I will send along three of my finest men."

"Three." McGee nodded impassively. Given his but recent induction into the General's company, much less his confidence, he did not think it remotely advised to betray frustration. Into enemy territory, arrayed against an army of the walking dead and a vile, shapeless offspring of the Great Goat Yossaguatha, the greatest general of the union would send four riflemen and a squadron of shiftless magicians? It was an absurdity he barely dared gaze at sidelong.

"What of young Lilienthal?" McGee inquired, grasping at straws. The spindly German was surely no use in open combat, but he had shown himself a companion with other valuable skills—even without his now-crippled flying frame.

"The inventor's son will be heading to his father's workshop, under guard by yet more men I can hardly spare," Grant replied, puffing impatiently out of his whiskers. "We hope he can repair his wooden bird and make it once again a strength to the Union."

At this, Phillips loosed an abrasive guffaw. "Ha! Such pretention. To take wing like a bird, as if nature were a thing free from fate." He turned again on McGee, stabbing a finger into the air, his eyes smoldering and his mouth a tight line. "Certainly you as much as me understand that reason cannot tame nature. She is not your tool, but a dark beast, imperfect and needful."

McGee found himself recoiling protectively, like a child suddenly confronted by a snake. "Surely I have no idea what you mean," he replied defensively.

Phillips drew closer still, grasping McGee by the shoulder. The magician whispered then, too low and quick for Grant to hear: "I know what you are."

McGee shuddered invisibly, but said nothing. He could not betray his panic to the General.

―

They were to set off almost immediately—the dark army was unlikely to wait. But as he had no things to gather nor camp to strike, Harry spent the time awkwardly swinging through camp on his new crutch and foot. The leather-and-wood contraption was awkward and painful, pinching and pressing and slipping—but he was grateful for it, a new innovation that Grants' field medics had on hand purely by luck.

Aside from trying out his new limb, he tried to track down Lilienthal. When McGee spotted him, he was directing the crew of men gathering the remains of his bird into a wagon. There was a strange new authority about him—the spindly man-child seemed to have embraced his sudden entrance into the theatre of combat.

"Aech, Meine Lieutenant!" Lilienthal embraced McGee about the shoulders with some semblance of manly camaraderie. They were equally filthy, so no harm was done in the exchange. Lilienthal then pressed the older man to arms length as if to administer a fatherly examination. "It has been such a magnificence to meet you. I do wish we had time to spare for chatter, but I will attend the machines." Between directing his own detachment and his strangely affected good manners, Lilienthal seemed to have aged five years in a matter of minutes.

McGee brushed off the boy's hands, suppressing the wry humor of the awkward ministrations to put on a stern demeanor. "Listen, Lilienthal, you are in this army now, like it or not. Grant doesn't seem eager to let you loose to putter about in your workshop."

The spry light of Lilienthal's face did not dim. "Of course he does not!" he replied, gesturing emphatically with an index finger. "No more of this puttering! Now, I will serve my country!"

McGee's eyes drooped mournfully. "You know nothing of war. Were it up to me, I'd see you far away from all of this, but I find myself in no position to excuse you."

Lilienthal smiled somewhat vacantly, like a pup expecting a treat. The import of McGee's words seemed to have missed him. McGee leaned in and whispered to him, grim-faced.

"Once you have reached your father's workshop, you must dispense with your minders and seek me out, below the line." McGee knew he was asking too much of the boy, but he was without options. "I do not trust these witches they send to aid me. I think General Grant may be, if not quite in their spell, then surely humanly afraid, and looking to their magic for reassurance."

Lilienthal nodded, now performing circumspection instead of bonhomie. "Tell me, Harry, what do you expect? What opposition?"

"We follow the dead and the horrible," Harry replied. "They're known for their weakness to nothing save light and fire."

Lilienthal nodded once, with the mock gravity of boy soldiers throughout history. The two shook hands and parted in silence.

As they set out, McGee found he could not well figure the value of his strange cohort. The tracking of the Confederate force was simplicity itself—the walkers and the Young together cut a swathe of darkness and death across the foothills and forests, a path wide and clear enough to be followed practically by feel. Failing that, the smell would be enough—char and rot, and some further mysterious unholiness. McGee located this last on the morning of the second day when, while briefly and tentatively out of saddle, he placed his good foot directly in it. There was a sense of falling and wetness, and when he looked down he saw his boot encased in a greenish orb of fat or saliva, strange and clear and half-solid. Leaning precariously against his horse, he pulled his boot off and shook it free of muck. High Witch Phillips, riding a few steps behind him, laughed.

"Ectoplasm," he observed. "Don't look so concerned, it's quite harmless."

McGee of course knew of the otherworldly material, and knew it was not in fact materially real. Nonetheless, he took a few moments to wipe it from his boot. The leather would still appear wet well into the next morning, when he would wake to find it inhabited by a half-dozen hand-sized tree-slugs.

The High Witch was more than generous in this and innumerable other lessons. His expertise, it seemed, extended from the curative properties of Appalachian mosses to the tactical calculus guiding the southern voudun.

"Of courses, there can be no doubt as to the path of this Youngling, but who knows where they might have sent the detachments of dead and living," he mused on the third day of their travel. "To my mind it seems that if they captured anything of real value, they would send it straight to Atlanta—not on this wandering amble abreast the Mississippi."

McGee had noted the same problem, and it more than worried him. One of the silent underling witches kept a detailed map by way of the stars, and the path of the Young, followed at first out of nothing but expedient certainty, had turned subtly away from Atlanta, and towards Louisiana—perhaps even New Orleans. McGee's presumption had been that, if they indeed were chasing some component of a new weapon, it must be headed for Atlanta, where the southron's workshops made what feeble attempts at innovation that they could.

There was no way of knowing now, though, whether the Young and the rest of the force had been separated—they had seen no diverging path, and any sign of living or dead men with the Young was utterly effaced by the huge sign of the rolling abomination. And, of course, it was the rest of that force that would be carrying the vital item it was their quest to reclaim.

At least as worrisome was the possibility that they were being followed. This alert too came from Phillips' unsettling cohort, when a strong tailwind picked up on the fourth day. The speechless men sniffed the air like dogs, though their faces remained invisible beneath their filthy cowls. Phillips noted their unease, and leaned down from his horse as if to consult them. From only a dozen feet away, McGee could hear no words exchanged, but when Phillips once again rose in the saddle, he announced that some magical presence was behind them. McGee thought at once of his twisted brother, so close behind when they departed.

The High Witch's imperious manner and dismissive self-absorption fed the flame of McGee's quiet rage with each passing day—as did McGee's apprehension about just what Phillips knew of him, and how. But the newly-minted Lieutenant could not deny Phillips' expertise. When the matter of the enemy's tactics came up once again on the fifth day of their trek, McGee, through gritted teeth, probed a bit into the lore of the voudun.

Phillips sighed deeply, his habitual theatricality not quite masking a note of true melancholy. "It is a sad irony in some ways, but in the end I can only call it justice. For many hundreds of years, they were no different than us, really." He gestured to the masked shapes who shambled behind them, too hateful for any horse, as if contrasting hypothetical bloodthirsty brigands with a clutch of respectably plump burghers. "And surely, when the first of the Aro arrived in Chesapeake with their dark bondsmen for sale, they would have seemed but little different from any Irish wastrel offering his nephew for a houseboy."

He shook his head and lowered his gaze to the blackened earth at their feet. "But surely, surely it was not mere habit that drove the planters on in their folly. They saw the Aro, in their majestic tall ships, laden with golden ornaments, their harems and their weapons. They lusted for the powers that had made those black bastards lords of the East. They began to suspect that our old ways were mere cantrips and lightshows."

"Is it true, then, that the voudun outstrip the witches of the North?" McGee did not shy from the opportunity to needle Phillips a bit.

Instead of bristling, though, Phillips guffawed heartily. "It is said that a bird in the hand is worth two in the bush, but in this case the bush was empty all along. It was nigh a century ago now that the Aro first offered the Lord Witch Samuel Baker of Alabama initiation into the secrets of Ekpe—for a handsome price in blood and treasure, of course. For an entire century the fine families have been sending the fifth part of every harvest and the most beautiful of their servants' daughters back on those tall ships to Africa. All for the promise of power.

"It has won them nothing but strife—and a sad decline in fashion sense. I have faced many voudun in the field, and though their manner has all the menace of the hunger they have earned themselves, the only true art of Ekpe seems to be charlatanism. They all fall to true craft, the last scraps of which they are now forgetting. It has become clear to all with eyes to see that the ladder of Ekpe has no pinnacle."

Harry glanced down at his wooden foot, so much easier to forget as it perched lifelike in a stirrup. He had been maimed for life by what Phillips called charlatanism.

The conversation paused for a moment, and McGee contemplated the odd contrast before them—the blasted path winding down the center of a lush, green forest.

"If what you say is true," he asked, "Why does the south continue to pay this tribute? Why are we at war, if they gain so little from their vile practices?"

"Ah." The witch's eyebrows danced with the pleasure of a pedant. "They now rely entirely on the Aro for labor. Their industry has been stunted at the root by the trade, but now they cannot do without it. And while I would sooner submit to a firing squad than be in even the most abstract bondage to a nigger, the Southerners seem to gain a satisfaction from owning men that exceeds the pain of being owned in turn."

McGee's eyebrows arched at Phillips' venom. He'd known freedmen back home, one or two passing well. He was no abolitionist, but as the war accelerated he had less and less patience for such dismissiveness towards the entire black race—and was surprised to hear it from one fighting on the side of Union.

Phillips continued, gazing into the treetops. "But it is yet more insidious than that. Not just their economy, but their society has come to be structured by the Aro and their secrets. The handsigns of Ekpe are passcode to the finest sitting rooms in Birmingham, and respectable ladies wear the sigil of the bloody trident beneath their bodices. It is considered a disgrace for a man of means to rank lower than the fifth embrace, even if he lacks the knowledge to cast a dog's ghost from a baptismal." Phillips spat vehemently on the ground. "Meanwhile, boys die by the thousands, telling themselves comforting fictions of honor and homeland."

McGee was not unmoved. A rather vaguer version of Phillips' narrative was now part of Federal soldiers' training. The general public of the North as yet knew little of about the debt bonds between the Southern lords and the Aro Confederacy, though: President Lincoln, for whatever reason, had chosen to let what had been secret remain so. McGee wondered at this still, for surely the concrete threat of international influence would have enraged the North more widely and surely than the abstract moral peril of bondage. McGee thought again of his brother, the voudun who defensively clung to the misnomer of witch-commander, and wondered what he had granted to the Aro in exchange for his power. For the first time in some years, he thought of his blooded enemy with a hint of sympathy amidst the rage and sadness.

By the seventh day, McGee's sextant showed them to be somewhere at the far eastern edge of Memphis. He was taken somewhat aback by this finding, but attributed it to the ease of following such a well-cleared trail. It was clear, moreover, that they were beginning to close with their quarry. McGee spotted a deer sprawled awkwardly at the edge of the bleak trail, and when he gingerly dismounted to get a closer look, he found that it was still breathing. Blood frothed from its mouth and its eyes showed the blankness of the final transition, but its wound was so grievous it could not have been inflicted more than an hour before. A long and horrific rend ran almost from its panting mouth down its twitching flank, with ragged, wandering edges that no sword had left. Leaning precariously over the creature, McGee saw that the wound was already coagulating strangely, blood turning to deep black at the edges, almost no crimson escaping from the fleshy declivity to run down into the loamy dirt.

"I would stay clear of that, were I you." Phillips gave the advice almost lazily, his voice bemused or indifferent, from where he sat on horseback some yards away.

McGee was about to ask why—and in the next instant he saw. The black blood moved in strange, irregular, pulseless waves that had no relation to the deer's ragged breathing. Tiny nodules formed and receded, and shades of color drifted just beneath the reflective black. McGee thought of the ectoplasmic ichor he had stepped in, and of the slugs that had found it so entrancing. This, he sensed, was not the settling of death, but the emergence of some obscene new life.

Seeing McGee's dawning recognition and revulsion, Phillips laughed his haughty laugh and turned to continue down the path. The faceless wizards of his train followed behind, all emitting a wet, birdlike clucking. McGee shuddered when he realized this was their echo of Phillips' laughter.

"Jackson!" McGee called. Of the three men General Grant had deigned to spare for the mission, Jackson was the most competent. His reassuringly straight bearing, square head, and neatly trimmed moustache concealed a rock-like stupidity, but he was strong and dogged enough to be entrusted with the small fire-cannon strapped to his dull-grey mount.

"Burn this thing immediately," McGee commanded, gesturing to the deer as the fireman rode up. The strange convulsions of its blood were invisible from a distance, and taking Jackson for none too observant, he felt no need to explain his order.

"Yes sir," replied the stolid trooper. As McGee haltingly remounted his horse, Jackson turned his own gelding broadside to the deer, then twisted around to give several preliminary pumps to the Gesner box. This done, he unholstered the wand, careful not to tangle its tubing, and used the complicated self-igniter to set the wick at its tip aflame. He then gave aim to the corpse from roughly ten feet away, and depressed the long lever that acted as a trigger.

Fire leapt in a liquid arc, connecting squarely with the deer's tainted corpse. When the fire hit, though, long black strands splashed in all directions from the deer's body, as if a heavy stone had been thrown into a dark pool. McGee sensed something of desperate escape about this explosion, and a few of the strands arced far enough in the direction of Jackson that the slow soldier twitched slightly in surprise. The line of kerosene swept momentarily away from the deer and its strange inhabitant, setting ablaze a small pine just next to the carcass. It was not clear to McGee whether it was the tree or the black thing that then began emitting a quiet, hissing scream.

After a few more moments of focused fire, there was nothing recognizable of the deer or its passenger—only a blackened, flaming lump without the honor of a pyre. Jackson doused his wick, holstered the wand, released the remaining pressure from the Gesner box, and turned his horse back towards McGee. He was gently shaking his shooting hand, and examining it with concern.

"If you know what that damned thing was, sir, I'm curious. Stings like a dickens." McGee took a look at the hand, and saw a slight red welt. It seemed that though most of the panicked tendrils had fallen short or been repulsed by the flame, one had succeeded in brushing Jackson before the fire set it to rout.

McGee hoped it wasn't serious. "We'll have the witch look at it. He'll know what to do." McGee, Jackson, and the other two Union men moved to catch up with Phillips and his cadre, who had kept moving, disinterested in the show.

"Phillips," McGee shouted ahead, "If you knew that thing was dangerous, why didn't you warn us?"

Phillips hied his horse to a stop, his silent sub-witches pausing as well. "Dangerous? That thing?" He laughed his affected, aristocratic laugh again, apparently for the benefit of an invisible theatre audience. "Why, it's just a little one, and you're off to fight its mammy."

"Wait," puzzled McGee, momentarily distracted from the immediate worry. "That thing was an… offspring?"

A certain wistfulness ran through Phillips' reply. "No, not really… more's the pity. The Young who passed this way cannot reproduce, per se. It is not properly a living being—only a corporeal manifestation. Think of it as a sound, which can leave its own echoes, but whose ultimate author is invisible." At that, he made an obscure gesture towards the sky.

"Well, get or mare, the thing stung one of my men." Harry waved Jackson up from behind him, and the dour man held out his hand, as if it were evidence of a schoolyard scrap presented to a headmaster.

"Oh dear," clucked Phillips, gazing down his nose. "That will have to be amputated, I'm afraid." And with that he began to turn back up the blighted row, toward their quarry.

As Jackson sat silently, apparently stunned, McGee moved swiftly after the witch, brusquely squeezing his mount amongst the shuffling witch-drones that clustered around their commander. "What in God's name do you mean, amputated? It's just a little scratch!"

Phillips didn't pull his horse up, and barely turned his head in reply. "Suit yourself. I'm sure your judgment of supernatural wounds is as well-informed as mine." He gave an indifferent wave over his shoulder as he continued.

Harry was just about to turn back to confer with Jackson, when he suddenly heard a distant crashing sound from the wood ahead and to the right. Both he and Phillips pulled up their mounts, and the witch corps perked their heads like attentive pets. The sound, of cracking twigs and scraping leaves and heavy thumps, grew louder by the moment, the sound of something running almost directly towards them. Without speaking, McGee and Phillips each began to move across the swathe, putting its black expanse between themselves and whatever they awaited. Phillips' faceless witches and McGee's handful of troops all followed. In the next moment, a wild baying joined the sounds of splintering wood and panicked flight—dogs, or perhaps wolves.

If whoever were following them had caught up, Harry was sure they would have made a more subtle attack. He thought of his brother, and wondered whether he could trust Phillips' confidence. Harry and his troops pulled out weapons, while Phillips and his creatures readied themselves for attack in ways more subtle—in just a few moments, an observer would have seen the air around them turn strangely translucent and distorted, as if a lazy summer heat had descended into the cool woods.

Finally, the sounds were nearly upon them. McGee raised the long barrel of his Army revolver, balanced it across his upraised forearm, and sought to draw a bead on whatever he could see—but the wood was too thick, right up until the last second. Then a man appeared, running wildly—only to tumble in the next second, as the edge of the Young's blackened trail dropped out from under his feet. There was a tangle of arms, legs, an orange shirt, swearing—and in the next moment the figure leaped up and made to continue its frantic flight.

"Halt!" shouted McGee, twitching the Colt emphatically. He had surmised the situation—for the man's skin was black as coal, his hair a series of tight knots, and his manner unmistakably that of one who must escape, or die in the attempt.

The runaway slave looked at McGee and his pistol, and stood in stunned terror. He raised his hands above his head, his eyes frantic but his body carefully restrained. He seemed unable to decide whether he preferred his new situation to the one he'd left behind.

Phillips had understood the situation as well as McGee—but where the lieutenant remained on guard, the witch emphatically waved the escapee towards them. "Come over here," he said. "We will protect you."

The baying—now clearly recognizable as hounds on the hunt—was growing by the moment, and the slave did not need much more persuading. Now ignoring McGee's gun, he bounded lithely towards Phillips and his ghastly cohort, then nodded curtly at the High Wizard. "Them that's after me ain't got much mercy, so I'll take any you can spare."

"Phillips, what in God's name are you about?" McGee had lowered his pistol, and now turned to face the conjurer. "What can we gain from interfering in such a dispute?"

"Oh, we can gain much, my practical friend." Phillips replied while keeping his eyes firmly directed towards the treeline. The howls were now a fierce storm, and in the next second a pack of five or six hounds burst over the lip of the monster's path and into the open, their flapping ears and sad eyes betraying little of the brutality of their errand. Before they had come a handful of steps, though, the dogs shrank, scrambling, suddenly whimpering, their eyes flickering to the shrouded witches that surrounded Phillips. "For one," offered the smiling witch-commander, "A bit of entertainment."

Just as it seemed the dogs would fall back in a panicked mass, Phillips' bleak troupe raised their hands in unison their bodies tense with focus. The animals froze in their tracks. Their moist eyes went blank, seeming to stare off into the long distance. It seemed perhaps they were listening very intently.

From beneath the witches' cowls, there came a sudden keening, a kind of atonal whistling whine, airy yet sharp. A crackling blue mist drifted from their sleeves, moving swiftly along the blackened ground until it reached the paralyzed dogs.

The dogs' stiffness mounted now, their muscles visibly tense. They trembled lightly, then began to shake. One shat itself violently, the contents of its intestines voiding like a popped blemish, and the rest followed in turn.

"Good God!" Harry coughed. He was terrified and rapt, suddenly unable to turn away or interrupt the unfolding horror. He noted blankly that one of the dogs, with large brown circles over its haunches and forehead, bore an uncanny resemblance to a pup he'd played with as a child.

Then they fell, one after another, unable to stand as their legs twisted upon themselves, their bones cracking under the force of their own spasmodic muscles. They bled from their noses and mouths as their jaws clenched with more ferocity than they would have reserved for their most hated quarry. Their eyes finally clenched shut, and something thicker than tears trickled out to wet the dirt as they choked on unimaginable pain. Harry's stomach heaved, and he heard Jackson retching violently behind him.

The half-dozen bodies soon went still and silent. They were no longer dogs, but fresh remains, their bones and flesh rent no less thoroughly than by a great bear. Harry heaved a ragged sob, his resentment for the witch-leader suddenly hardened within him. He heard the escapee breathing with as much miserable panic as if he were still being chased—now unsure if this were the frying pan or the fire.

In the next moment, the dogs' master stumbled into the clearing—a gaunt, worn white man, his broad-brimmed hat streaked with mud or dust. He pulled up short and crouched defensively, jerking a blunt shotgun to and fro, his confusion turning to horror as he took in the welter of mutilated corpses. Instinctively, he fixed the long piece's sights on Phillips. Though both of the gun's percussion hammers arced backwards like venomous snakes, Phillips sat bemusedly on his horse, far too relaxed, leaning forward with his wrists crossed on the pommel of his saddle.

"I'm afraid my friends don't much care for dogs," Phillips offered, seemingly by way of apology. His minions again made those reptilian laughing sounds, with echoes of demonic hatred and creeping sickness that sent chills up Harry's spine.

The white man's eyes next lit on the slave he had been chasing, who now stood beneath the protective girth of Phillips' horse. The sight seemed to give him some strength, if only by offering a recognizable connection to reality.

"Josiah!" he yelled. "You can do no better than to take up with Yankee witches?"

Harry noted that the black man was about to speak when Phillips replied on his behalf. "From what I see, he'd be hard pressed to find worse company than yours. But perhaps I mistake the situation?"

"The situation's not yours to mistake, white hat, and the chattel's not yours to interfere with! Stand down or I'll blow your damned Satanic head off!" The man was winded and panicked, and his weapon wavered in his hands. No more sound came from the woods behind him—he alone had undertaken the chase.

"You'll not want to be pointing that thing the wrong way," Phillips intoned. McGee saw one of Phillips' shrouded under-witches move toward the slaver with strange, sidelong steps. There was nothing subtle about them, but the white man seemed to make no note of the aggressive move, remaining entirely fixated on Phillips.

"You release my property! I paid the Aro for him, square dealing." A strange clip entered the man's tone—perhaps the stiffness of his mounting fear. "War spoils, fresh from Africa."

The slave Josiah gave a tense laugh, the tentative joy of a man who saw reprieve at hand. "This fool thinks I come from Africa? He must be deaf. I was born in Boston, and I aim to get back there." McGee did note a bit of Massachusetts in the man's speech. But it didn't make much difference—he misliked the way things were unfolding, and placed no trust in the witches' judgment.

McGee raised the Colt and squeezed off a round that kicked up dirt near the feet of the creeping witch. "Stop this!" he shouted. The shrouded figure pulled up sharply, only halfway to its quarry—and the slaver, seemingly snapping back to reality, stared in confused horror at the cowled cultist, as if it had appeared out of thin air. His shotgun again dipped and swung, making McGee flinch.

McGee turned to Phillips. "You've had your fun. I'll not see you treat this man to the fate you gave his dogs." Then he addressed the frightened, angry rustic, all the while training his pistol squarely on the man's soiled hatbrim. "But I likewise doubt my allies here are in any mood to let you leave with your slave. You'll just have to do without him."

The man gazed around at the dozen-odd dangerous figures who opposed him, and finally seemed to shrink. The shotgun's threatening depths finally descended to the ground. His voice turned piteous and pleading. "You Yankee cunt. Without Josiah, I'll never keep body and soul together. As if I'm doing so well with him…" He glanced at the dirt-streaked homespun of his shirt and pants. He gazed on the remains of his hunting dogs, emitting an involuntary whimper.

"Drop your weapon," McGee demanded firmly. "You are outmatched. I guarantee your life if you go in peace."

The steam had gone out of the slaver. His dogs were dead. There was no denying the gravity of his situation. He raised his eyebrow at the witch that had moved towards him, and towards the mounted Phillips.

"Call back your man, Phillips." McGee spoke sternly.

"He's not my man," Phillips replied, "But purely his own. And not unlike myself, he hates slavemasters even more than he hates dogs." But despite the menacing words, the hunched white shape gave something like a shrug and returned to his cohort.

The man who but moments before had been hunting his escaped chattel now slowly crouched to place his weapon on the ground, with no choice but to place faith in a squadron of Yankees he instinctively loathed.

"When we're gone from here, you can come back for your weapon." McGee offered. If nothing else, he'd not leave the man without a means to feed himself.

The onetime pursuer rose and backed cautiously away from those who had ruined him. His shoulders were slumped, and McGee now saw the deep lines carved in his face. McGee watched as the man took small, fearful backwards steps, almost to the edge of the wood. McGee was sure this was just one more hard chapter of a hard life.

"I'm very sorry about your dogs," McGee offered, almost involuntarily.

"Don't be," replied the white man, some hint of defiance returning with each steps away from the Northerners. "You'll look no better than them when our boys catch you." With that, he spat a wet gob into the dark earth, and turned into the woods at a run.

The slave Josiah covered the thirty yards between the witch-commander and his former master's shotgun in the wisp of a breath. He snatched the weapon from the ground, then skidded to a halt and swung it to his shoulder, barely feet from the white man. McGee, taken fully by surprise, had no time to react. The hunter, turned quarry, had just enough time to halt in his tracks and turn halfway around, seeking the source of the sudden sounds. He found only the gaping maw of his own weapon, and McGee flinched away from the click and boom of wild vengeance.

But when Harry opened his eyes, he did not find gore and death. The slave Josiah panted wildly, cradling the discharged shotgun—and before him, his former master was strung amongst the trees like a puppet. A mesh of uncountable fine lines tangled through branches, and the man's shredded clothes hung in further tatters. Through the rags McGee could see tiny welts, red as a hundred carnelians, where the lines clung to his skin. The shot from below had carried him a foot or two off the ground, where he now hung, affecting a ragged laugh.

"Josiah… you never could control that temper of yours. Not even long enough to reckon I'd be coming after you with netshot."

But again the black man moved with almost supernatural speed, silencing the laughter with the butt of the shotgun laid across the captive slavemaster's cheek. McGee could have sworn he heard crunching bone, and as the captive man's head drooped, blood gushed from every hole in his face.

"Stop this!" McGee shouted, spurring his mount across the dead sward. "I gave that man his freedom, and I'll see it honored."

The slave turned and spat towards McGee. "Ah yes, you Yankee Doodles love to *give* freedom. But not justice, never that." He turned again and swung the shotgun, holding it by the barrel and arcing it like a mining pick for his former master's skull. But McGee was already into the wood, and by some miracle of reckless balance managed to awkwardly sprawl from the saddle onto the freed slave. He deflected the deadly blow, which nonetheless glanced into the bleeding man's arm, audibly breaking it as McGee and Josiah fell in a heap.

"You stupid bastard!" the slave practically screamed as he pushed against McGee. With only one good foot, McGee knew he'd be hard-pressed to stand up, much less fight, so he struggled merely to pin his opponent, holding down his arms.

"I'm doing you a favor," the slave growled as he struggled. "It's obvious you're headed south, and with only ten of you, I'm sure–" He grunted, levering an elbow under McGee's faltering grip. "I'm sure your mission is secret. This one would have drummed up every two-bit mud-farmer's militia he could find." Even as he talked, Josaiah shifted his strength with finesse, quickly pushing McGee off of him. The Union man, knowing himself outmatched, sat in resignation as Josiah stood and finished his address.

"Let him go, and you'll have a hundred toothless Rebs crawling up your ass with rusty muskets in half a day." He panted and brushed dirt from his tunic.

"Alright then," rasped McGee from his undignified sprawl, "Kill him if you wish. I've clearly got no power to stop you." But McGee spoke half in jest—unremarkable though they were, his tiny retinue of soldiers had at least had the sense to move up and support him, rifles raised. Josiah saw them, and made no further move for the shotgun.

"We'll leave the slaver as he is," Phillips declared as he casually rode up on the scene. "Perhaps an ally will find him. Or perhaps he'll learn something of endurance." He smiled, clearly relishing the latter possibility.

Josiah laughed too, and, glancing almost casually at the guns trained on him, walked past McGee to address Phillips. "What exactly are you doing here, witch?"

"We're here to free a slave far more important than you," Phillips replied.

"We seek a weapon," McGee chimed in, annoyed that the freed slave seemed to think Phillips the leader of their troupe. "The rebels captured it from a storehouse a week's ride north, and we think they're taking it to Atlanta." As he spoke, he used a tree to pull himself upright, then wobbled towards his horse, barely staying upright without his crutch.

Josiah looked up and down the burned path, then back to McGee. "You're following this trail?"

Phillips replied for him. "The rebels use one of the Young for their own perverse ends."

"If you wish to pay back your freedom, you'd be welcome to join us." McGee, still smarting from the freedman's victory, could nonetheless see he'd be an asset.

Josiah laughed. "Why should I go south with you, into more damned danger, when I could move north and keep both my freedom and my head?"

"I don't treasure your company, African," Phillips said, "But your freedom is surely safer with us. There's a voudun harrying us from the North who'd turn you into an even worse kind of slave. And have no fear, we'll be quit of you as soon as the opportunity arises."

"Fine," Josiah huffed. "I'll travel with you. I suppose I do owe you something." He glanced back at his former owner, who still hung, bleeding and now insensate, in the mess of hooks and lines. "And I'm content hoping this one starves to death."

The slave kept the shotgun, insisting it was his as payment for service to his former master. He also took the ammunition pouch from the bleeding man's waist, then eagerly rifled through it, searching for a more deadly pair of paper cartridges. They gave the former slave one of the spare horses, which he rode bareback with some skill. He in turn offered his full name as Josiah Mayweather. McGee pressed him for his story—a stunning one, it turned out, of a free black youth stolen from his Northern home into an adulthood of bondage. In Josiah's eyes, the Fugitive Slave Laws had left little difference between Northerners and Southerners. Even now, free, with war underway and Emancipation on every man's lips, he seemed to have little hope of a happy future in the United States. No wonder, thought McGee, that he was so entranced with violence. He surely saw no other way.

The men did not have long to learn each others' stories, though. Early the next morning, before the dew was off the trees, they saw the end of the black trail they had been following. Trees rose up behind a gaping hole in the earth, as big around as a barn, with its depths revealing nothing.

"She's gone to ground," declared Phillips, taking in the scene. "We've no choice but to go in after her. And we're sure to find our captured weapon there, too." McGee heard something odd in the tone of this last part, as if it were a mere afterthought.

"To go into such a place seems foolhardy, to say the least," observed McGee. Mayweather nodded agreement.

"Of course, I'd surely prefer the situation were different," Phillips offered in a condescending tone. "But if we hesitate, they'll only become more of a threat. They're likely fortifying their position right now."

"Wait, though," asked Mayweather. "If they're trying to get this weapon of yours to Atlanta, why would they be digging into one spot?"

Phillips didn't skip a beat. "For the Young, space does not conform to the rules you and I know. Given some time and security, she'll be able to move herself and her retinue anywhere, far faster than we could hope to keep up. We may only have a matter of hours to interrupt her passage. Already, she'll be creating defenses from the substance of her own body."

McGee suddenly mistrusted Phillips. But the trail stopped here—they were too committed to this path to turn back. If the weapon were anywhere else, they were already lost.

He commanded Jackson to remove the flame cannon from his horse and carry it. The other two soldiers he would leave to guard their backs. McGee was loathe to leave the horse that had limited the inconvenience of his hobbled foot, but he had no choice—Phillips insisted the tunnels would get narrow. He pulled the stout wooden crutch from across his saddle.

So arrayed, the two soldiers, the freedman, and the half-dozen witches, each bearing torches of rag and kerosene, stood before the gaping, fresh-made cave. As they contemplated the dark path before them, McGee noted a smell wafting from the hole. It was like nothing he'd known before—a smell not of death or rotting, but of strange life, salty and lush and metallic. There was something else, too—something he could neither smell, nor hear, nor see, but only feel—a vibration against his skin, a low thrum that pulsed uncomfortably at his very core.

"What is that… rhythm?" He asked, his gaze not wavering from the blackness into which he must descend.

When Howard Phillips answered, it was in a truly somber tone, ripe with real fear.

"That," said the grim magician, "is the heartbeat of a god in chains." ✤

To be concluded…

David Z. Morris lives in Tampa, Florida, where he writes essays and fiction, and helps lead The Venture Compound, a nonprofit community space devoted to avant-garde and outsider art and music. You can find more of his work, including more fiction set in the world of "The Scouts of the Pyre," at davidzmorris.com.

JOIN THE WAR

The Red Fork Empire

The Red Fork Empire (RFE) is a collective of people who want to express themselves creatively. Not necessarily as professionals, but because using your imagination and being creative is fundamental to truly enjoying life.

To help spread this message the emperor and the citizens perform as their personas, speak on event panels, run Imperial games, volunteer at galleries and museums, curate as well as participate in art shows, and contribute creative works in the name of the Empire. We speak about how the arts program is necessary in schools, that it is just as important as the sports program if not more so. We spread the message to everyone we meet, to encourage those who may not think that they are artists to accept that they are, and they have an artistic voice that needs to be expressed.

The Red Fork Empire helps its citizens with their creative endeavors. We offer advice, critique, collaboration, and help as well as aid in promotion, to get the word out about the current project you are working on. We all want to help each other's creative pursuits.

The enemy of the Empire is the *dull*. The dull is what stops the creative process flat. It is the thing that keeps people from realizing they can still play. The dull infects and possesses, and it inhabits the uninspired and reluctant. For those infected by the dull, there was a time when you imagined ridiculous things, yet at some point in your life it all went away. The dull took hold and you think you can't play anymore. The RFE says you are wrong. You can create, you can draw a line. You can imagine something and express yourself. It is the main focus of the RFE to combat the dull and to remove it from everyone.

It doesn't matter if you are good at what you do—as long as you are trying, that is all that matters. There will always be professionals, but you don't need to give up walking just because someone else can run.

The other aspect of the RFE is that the emperor is the ruler of the multiverse. All of the citizens of the Empire help the emperor in his fight against the dull. The home world of the RFE is New Byzantium and its capitol city is the Citadel. It is in this multiverse that the emperor encourages people to play. Sometimes people need a push or something to get them started. The Empire and the multiverse is that springboard.

The Empire encourages citizens to contribute creativity and imagination to it, to help its story and mythos grow. In this way, it is the people who build the Empire and shape what it can be. When a citizen creates something in the name of the Empire, they receive an armband. When they do more, they receive medals. These are awarded so as to encourage creativity and imagination, to encourage having fun. By being creative we learn more about ourselves and who we want to be. In using our imagination, we find another way to communicate with those with whom there might not have otherwise been any connection.

The RFE will hold the line so that worlds of make believe live on and never are crushed by the dull.

Keep up the fight.
—*Emperor of the Red Fork Empire*

AGAINST THE DULL!

The Chaos Front

The Chaos Front is a loose alliance of individuals and organizations that wages a war against both imperialism and the dull.

We shall never bow to empire, nor to any tyrant no matter how benevolent or entertainingly mad.

In our ranks are pirates and anarchists and others, including all of those among you who wish to be free of the shackles of "citizenry."

Our guiding principles are:

- Opposition to the dull and banality.
- Opposition to all empire and oppression.
- Autonomy for all groups and individuals within the Front—no individual group within the Front shall control the others.
- Solidarity within the Front—we help one another in times of distress.

We are the Chaos Front, and you shall know us by the chaos we sow, the joy we proliferate, and the fun we have.
—an autonomous, anonymous member of The Chaos Front

Full disclosure: this magazine's publisher, Combustion Books, is allied with the Chaos Front. The magazine itself maintains a neutral stance between the two factions (though remains steadfast in its opposition to the dull) and many of our contributors and editors are citizens of the Red Fork Empire.

SUBMIT TO NO MASTER!
BUT CONSIDER CONTRIBUTING TO STEAMPUNK MAGAZINE!

Please keep in mind before submitting that we publish under Creative Commons licensing, which means that people will be free to reproduce and alter your work for noncommercial purposes. We are not currently a paying market. Please introduce yourself in your introduction letter: we like to know that we're working with actual people.

FICTION: We appreciate well-written, grammatically consistent fiction. That said, we are more interested in representing the underclasses and the exploited, rather than the exploiters. We have no interest in misogynistic or racist work. We will work with fiction of nearly any length, although works longer than six thousand words will be less likely to be accepted, as they may have to be split over multiple issues. We will always check with you before any changes are made to your work. Submissions can be in .RTF, .DOC, .DOCX, or .ODT format attached to email. Please include a wordcount of your story in your email. We receive more fiction submissions than the rest of the categories combined and reject a majority of what we receive. Submit to collective@steampunkmagazine.com

POETRY: We are not currently accepting poetry submissions.

NONFICTION: We run a variety of nonfiction, including pieces on contemporary steampunk culture and historical articles about 19th century rabble-rousers, Victorian oddities, and other items of interest. Submit to nonfiction@steampunkmagazine.com

ILLUSTRATION: We print the magazine in black and white, and attempt to keep illustrations as reproducible as possible. Ideally, you will contact us, including a link to your work, and we will add you to our list of interested illustrators. Any submissions need to be of high resolution (300dpi or higher), and preferably in .TIFF format. Submit to art@steampunkmagazine.com

PHOTOGRAPHY: We do not currently run photography.

HOW-TOS: We are always looking for people who have mad scientist skills to share. We are interested in nearly every form of DIY, although engineering, crafts, and fashion are particularly dear to us. Submit to howtos@steampunkmagazine.com

COMICS: We would love to run more comics. Contact us at art@steampunkmagazine.com

REVIEWS: We run reviews of books, movies, zines, music, etc. on our website. We are looking for reviewers as well: please contact us if you are interested at reviews@steampunkmagazine.com

FASHION: Although we are quite interested in steampunk fashion, we are more interested by DIY skill-sharing than exhibition of existing work. If you want to share patterns or tips for clothing, hair, or accessories, then please let us know at howtos@steampunkmagazine.com

INTERVIEWS: Most of the interviews we run are conducted by staff of the magazine, but we do occasionally accept interviews conducted by others. However, it would be best to contact us before going through the work of conducting the interview to see if we would be interested in running an interview with your subject. collective@steampunkmagazine.com

ADVERTISEMENTS: We do not run paid advertisements on our site or in our magazine. We rarely run press releases. If you have a product you want our readers to be aware of, your best bet is to write a how-to article explaining how to make it, submit the media for review, or request an interview.

OTHER: Surprise us! We're nicer people than we sound! collective@steampunkmagazine.com

---

SteamPunk Magazine is published by Combustion Books, a worker-run genre fiction publisher based out of New York City. Issue #9 was released in May, 2013.

EDITOR: Margaret Killjoy

ART EDITOR: Juan Navarro

CONTRIBUTING EDITORS: Amy, Sarah Tops

CONTRIBUTING WRITERS: Larry Amyett, Jr.; Professor A. Calamity; Katie Casey; James H. Carrott; The Catastrophone Orchestra; Joanna Church; E.M. Johnson; David Z. Morris; Jamie Murray; Professor Offlogic; reginazabo; John Reppion; Miriam Rosenberg Roček; Kendra Saunders; Erin Searles; Douglas Summers-Stay; Charlotte Whatley

CONTRIBUTING ILLUSTRATORS: Benjamin Bagenski; Micheal Barnes; Tina Back; Sarah Dungan; Sidney Eileen; Doctor Geof; Allison M. Healy; Kelley Hensing; E.M. Johnson; Chaz Kemp; Cécile Matthey; William Petty; Geneviève Lamarche Reid; Sergei Tuterov; Charlotte Whatley

FONTS: Tw Cen; Adobe Garamond Pro; P22 VICTORIAN GOTHIC; Ehmcke; LAUDANUM

Made in the USA
Charleston, SC
19 July 2014